Tensegrity Structures Design Methods

Tensegrity structures are prestressed systems of cables and bars in which no bar is connected to the other and the structure has no continuous rigid skeleton. This general introduction presents an original general method for the design of tensegrity structures, the first configurations of which were found by trial and error.

The book begins with two-dimensional tensegrity structures, particularly tensegrity nets, tensegrity chains, tensegrity rings and tensegrity arches. These are then developed to original configurations of spatial tensegrity structures such as tensegrity slabs, primitive spatial tensegrity arches and primitive tensegrity domes, as well as more elaborate spatial tensegrity structures such as tensegrity cylindrical shells, slim tensegrity domes, tensegrity vaults and tensegrity caps.

- Presents a robust new approach to the design of tensegrity structures
- Extends tensegrity structures to new three-dimensional configurations

Tensegrity Structures Design Methods suits structural, civil and mechanical engineers and architects, as well as graduate students.

Oren Vilnay is Professor Emeritus and was founder and head of the Department of Structural Engineering at Ben Gurion University Israel. He is also former head of the Structural Engineering Section at Technion—Israel Institute of Technology.

Leon Chernin is Lecturer at the University of Dundee. He was granted a PhD in Structural Engineering from the Technion—Israel Institute of Technology. His research activities encompass both physical testing and numerical modelling.

Margi Vilnay is Senior Lecturer at the University of Dundee. She was granted a PhD in Structural Engineering from Heriot-Watt University. She is a chartered member of the Institution of Civil Engineers and the first woman to be elected Vice-Chair of COMEC (Council of Military Education Committees).

Tensegrity Structures
Design Methods

Oren Vilnay, Leon Chernin
and Margi Vilnay

CRC Press
Taylor & Francis Group
Boca Raton London New York

CRC Press is an imprint of the
Taylor & Francis Group, an **informa** business

Cover image: Oren Vilnay

First edition published 2024
by CRC Press
6000 Broken Sound Parkway NW, Suite 300, Boca Raton, FL 33487-2742

and by CRC Press
4 Park Square, Milton Park, Abingdon, Oxon, OX14 4RN

CRC Press is an imprint of Taylor & Francis Group, LLC

© 2024 Oren Vilnay, Leon Chernin and Margi Vilnay

ISBN: 978-1-032-44035-4 (hbk)
ISBN: 978-1-032-44036-1 (pbk)
ISBN: 978-1-003-37009-3 (ebk)

DOI: 10.1201/9781003370093

Typeset in Sabon
by Apex CoVantage, LLC

Contents

Preface

A tensegrity structure consists of a cable net and bars. The bars are connected at both ends to the net nodes. No bar touches the other and the tensegrity structure has no rigid skeleton. The structure gains its rigidity by prestressing forces.

Because the bars are much thicker than the very thin cables which are hardly visible, the tensegrity structure creates the illusion of bars floating in space.

At a very early stage, I was fascinated by tensegrity structures. I studied the conditions a tensegrity structure should satisfy and formulate the deformation of a tensegrity structure due to external load as well as its dynamic behaviour. These findings were published in a book (Vilnay 1990).

Since then, I continued to study tensegrity structures. In spite of research in other fields of structural engineering and university bureaucracy, I always manage to find the time to study these allusive structures.

Members of my family who are structural engineers were fascinated by tensegrity structures as well. Their contribution to this book is indispensable and cannot be ignored. They put an extensive amount of work to the publication of this book in its present form.

This book is devoted to the design of tensegrity structures and present different methods that can be used to determine the configuration of these fascinating structures. Various types of tensegrity structures are considered.

The tensegrity structures presented in this book are simple in order to demonstrate the principles of design as simply and clearly as possible. The reader is encouraged to follow these footsteps and to design more elaborate and intricate tensegrity structures.

O.V.

Introduction

The simplest tensegrity structure is a simple kite. The kite is a two-dimensional structure composed of two bars which are not connected to each other and a cable along the parameter. Children and adults have been playing with it for many centuries.

While studying in the late 1940s at Black Mountain College in North Carolina, the student Kenneth Snelson, while playing with novel structure sculptures, constructed a three-dimensional, the so-called X-Module, tensegrity structure. The X-module can be seen as two kites one on top of the other. Richard Buckminster Fuller, a visiting professor, realized the novelty of this structure and by combining the terms of tension and integrity coined the term tensegrity to define these structures.

Snelson, who became a contemporary sculptor, elaborated this structure and constructed many tensegrity sculptures. The best known is the 18-meter high needle tower in Hirshhorn museum Washington, DC.

Fuller promoted the term tensegrity with his cosmic and operational philosophy and defined it in a poetic way, islands of compression in an ocean of tension, or more precisely "A tensegrity system is established when a set of discontinuous compressive components of compression interact with a set of continuous tensile components to define a stable volume in space" (Fuller 1961).

Unfortunately, this definition of tensegrity is ambiguous and does not single out Snelson type structures. Instead of defining a class of Snelson type structure, many structures in which tensile elements are in abundance can be considered tensegrity structures: cable nets, suspension bridges, suspension roofs, stayed columns, stayed bridges, etc.; tensegrity is then in the eyes of the beholder.

Some may argue that the Olympic stadium in Munich, Germany; the La Plata stadium in Argentina; the Kurilpa Bridge in Brisbane, Australia; the KSPO dome in Seoul, South Korea; and the Centennial Olympic stadium in Atlanta, Georgia, are tensegrity structures.

Some feel uneasy and refer to some of these structures as hybrid tensegrity or structure in the spirit of tensegrity.

To single out Snelson type tensegrity structures, a more accurate and precise definition is often used.

Tensegrity structures are composed of a continuous tensile net (cables) and compression elements (bars) which are connected to the tensile net only and not to each other. They are constructed with very few compression elements attached to the foundations. They are prestressable: prestressing induces tension to the cables and compression to the bars.

In this way cable nets, suspension bridges, suspension roofs, stayed columns, and stayed bridges in which all compression elements are attached to the foundations are not defined as tensegrity structures. Only structures with no rigid continuous skeleton, what Snelson defines "floating compression", are tensegrity structures.

Classical tensegrity structures are used in robotic and biotensegrity where tensegrity structures are used to model the human skeleton and the live cell.

When introduced, tensegrity structures perplexed the profession and the academics. After long deliberation, it was established that tensegrity structures have a certain similarity to cable nets. Some tensegrity structures are indeterminate, but most of them belong to a very specific type of mechanism, the so-called infinitesimal mechanism with the following typical properties:

- Both have a specific configuration in which they are prestressable and prestressing induces tension to the tensile elements and compression to the compression elements.
- When the infinitesimal mechanism tensegrity structures are loaded by the so-called "fitted load", the nodal displacements are small and are in the order of magnitude of the elastic deformation of common structural engineering structures.
- In the case where the infinitesimal mechanism tensegrity structure is loaded with loads which are not "fitted loads", the tensegrity structure cannot sustain these loads in the prestressed configuration. There are large nodal displacements to achieve equilibrium at all nodes. These large displacements can be often spotted by the naked eye. In the case of dynamical loading, the small displacements are with high frequencies and the large displacements are with low frequencies.

The analysis of tensegrity structures and cable nets is presented in Vilnay's book (Vilnay 1990).

Designing tensegrity structures was proved to be far more difficult than establishing methods of analysis. First tensegrity structures were built by intuition and trial and error. They were in the shape of polyhedras where the bars occupied the enclosed space. There were attempts to formulate the regularity of tensegrity structures and to provide the designer with tools to design them.

Snelson showed primary weave cells and the equivalent basic tensegrity modules (Snelson 2012). Pugh distinguished between three basic design patterns according to which tensegrity structures can be constructed: a diamond pattern, a jigsaw pattern and a circuit pattern (Pugh 1976). Motro defines single prismatic cells of tensegrity and extend them to tensegrity polyhedras and to complex tensegrity assemblies and to the formation of "Hollow Rope" (Motro 2003). There were also attempts to join tensegrity polyhedras to each other to construct double-layered tensegrity slabs and shells (Hanaor 1992).

The major difficulty in the development of tensegrity structures lays in the fact that structural engineers are not familiar with the design of infinitesimal mechanism. Often concepts used in common structural engineering such as triangulation and indeterminate structures are wrongly applied to tensegrity structures. Structural engineers do not feel at ease and even terrified by spatial tensegrity structures. The fact that tensegrity structures did not capture the imagination of the structural engineering profession is partly due to this fact.

In this book, a novel approach to the design of tensegrity structures in the language of structural engineers is presented.

The first part of the book is dedicated to two-dimensional tensegrity structures: tensegrity nets, tensegrity arches and tensegrity rings.

In the second part of the book, the two-dimensional tensegrity structures elements are used to form spatial three-dimensional tensegrity structures: tensegrity slabs, tensegrity vaults, tensegrity cylindrical shells, slim tensegrity dome, tensegrity caps and tensegrity domes.

Schematic patterns of simple structures of each type are presented and the conditions for a feasible tensegrity structure are formulated.

In the third part of the book, it is shown how by adding cables to the prestressed tensegrity spatial structures and prestressing these additional cables appropriately, it is possible to change the tensegrity structure from an infinitesimal mechanism to common indeterminate structure. These tensegrity structures are defined as tensstable structures.

It is hoped that these schemes of simple tensegrity structures will encourage others to develop and construct elaborate tensegrity and tensstable structures.

Tensegrity: literature survey

Origin of tensegrity

It is possible to find the roots of defining tensegrity structures in the second half of the 19th century with the work of Maxwell (1864). In his monumental work, Maxwell defined a rule for the static determinacy of three-dimensional pin-jointed frames (trusses), stating that the frame is structurally stable when the number of members in the frame is equal to or larger than three times the number of pin joints reduced by the number of constraints (minimum six) introduced by the supporting conditions. If this rule is violated, the frame becomes an unstable mechanism. Maxwell anticipated, however, that his rule is not essential because, under certain conditions, a frame with fewer members than defined by the rule can still be stable. One such condition is the state of self-stress or prestress that stabilizes the inherent mechanisms within the structure, that is, makes them infinitesimal mechanism (Calladine 1978). In his work, Maxwell pointed about the possibility of tensegrity structure but did not present any spatial example of such structure.

It is widely accepted that Fuller, Emmerich and Kenneth Snelson were the three "fathers" of tensegrity structures who were granted patents in the early 1960s (Fuller 1962; Emmerich 1963; Snelson 1965). However, the historical origins of tensegrity structures can also be connected to Constructivism, one of the Russian Soviet avant-garde modern art movements born in 1915 and flourished during the 1920s and 1930s in the USSR and then Germany (the Bauhaus) and Holland (De Stijl). Also, the aesthetic movement of Art Deco borrowed from it. In art, Constructivism was characterized by austerity, geometrysm and laconism, while in architecture, it focused on the simplicity of structural forms and functionality and purpose of structural systems rather than on their decorative stylization and appearance. Using Constructivism philosophy, the Latvian-Soviet avant-garde artist Karl Ioganson (Kārlis Johannson) created between late 1919 and May 1921 nine spatial structures (Gough 1998, 2005). He called them "cold structures", emphasizing their rational, lucid and simple geometry, the efficient and economic use of materials as well as the minimum effort required for workmanship and construction, which was associated with hot welding, melting and molecular bonding. Ioganson displayed spatial structures in 1921 in the Constructivist gallery at the Second Spring Exhibitions of the Society of Young Artists (OBMOKhU) at the Institute of Artistic Culture (INKhUK) in Moscow. The structures are composed of cable net prestressed by supporting bars. In some cases, the bars are connected to each other, while in some the bars only overlap. One of those structures shown in Figure I.1 is a simple spatial tensegrity structure.

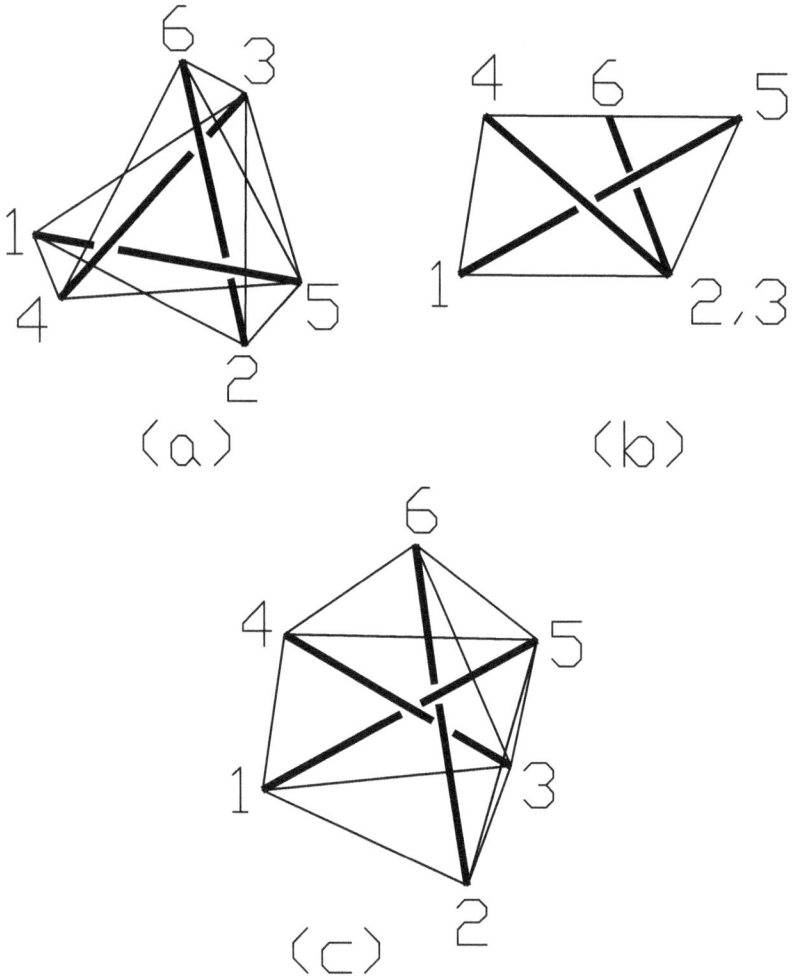

Figure I.1 Karl Ioganson's tensegrity structure: (a) plan, (b) side and (c) perspective view. This structure is also known as tensegrity triangular prism or Simplex.

Unfortunately, only photographs of the exhibition survived, for example, in works by Bela Uitz (1922), László Moholy-Nagy (1929), Christina Lodder (1992) and Maria Gough (1998, 2005). Since Constructivists considered a cross as "the minimum requirement for the existence of any structure" (Gough 1998), this principle was implemented in the creation of some of the structures presented by Karl Ioganson which are different variations of a spatial cross. Karl Ioganson summarized this principle as follows: "The construction of any cold structure in space, or any cold combination of rigid materials, is a Cross . . . right-angled . . . or acute- and obtuse-angled" (Gough 1998). However, further experimentation with stable structural configurations and different materials allowed Karl Ioganson to demonstrate that the cross was not an essential requirement for the existence of the "cold structure". This idea was implemented in the structure shown in Figure I.1, which lacks the central joint of the space cross.

The significance of Ioganson's invention was not understood at the time of its exhibition in the Constructivist gallery in 1921. Even today Ioganson is widely credited not as the inventor of the first tensegrity structure (e.g. Gough 1998, 2005; Burkhardt 2008; Skelton and de Oliveira 2009) but rather as the inventor of the proto-tensegrity Equilibrium structure (e.g. Emmerich 1988; Motro 1992, 2003; Gómez-Jáuregui 2009; Sultan 2009). It can be concluded that the reinvention of the tensegrity structures occurred mostly independently from Ioganson's work. David Emmerich did not presumably saw that tensegrity structure, since he refers in his work (Emmerich 1988) to Moholy-Nagy's book (Moholy-Nagy 1929) which includes a photograph of the Equilibrium structure and a photograph picturing only part of the Constructivist gallery at the Second Spring Exhibitions excluding the Ioganson's tensegrity structure. It is not clear whether Fuller was aware of Ioganson's works, while Snelson saw the Ioganson's Simplex much later in the 1992 catalogue of the Guggenheim show picturing Koleichuk's reconstructions of Ioganson's spatial constructions (Snelson 2003).

A breakthrough occurred at the Black Mountain College in North Carolina, where Fuller was a visiting professor during the summer of 1948 and the institute director in the summer of 1949. In the summer of 1948, a young artist, sculptor and photographer, Snelson came to Black Mountain College to study with a famous German-born artist and educator, Joseph Albers. It is important to note that during the 1920s and early 1930s, Albers studied and then taught (together with László Moholy-Nagy) at the German art school, Bauhaus, which was heavily influenced by Constructivism. At Black Mountain College, Snelson met Fuller and attended his classes. "[In the college], Fuller . . . spoke constantly of tensional integrity. Nature relies on continuous tension to embrace islanded compression elements, he mused; we must create a model of this structural principle" (Edmondson 1986). Furthermore, Joseph Albers spotted Snelson's talent for three-dimensional construction and asked Snelson to assist Fuller (then a new faculty member) in assembling models of spatial geometric structures used in the evening lecture to the college (Snelson 1990). Snelson was captivated by Albers' Bauhausian art and design ideas and Fuller's futuristic vision of spatial geometry and decided to become a sculptor. In the fall of 1948, Snelson returned to the University of Oregon, where he enrolled on engineering and began making a series of small structures. In three of those works, Snelson investigated modularization and equilibrium. The first two structures represented variations of a moving wire column made of similar modules stacked vertically. Each module included an arc-shaped wire element weighted with clay balls at the ends and balancing on a central swivel. Those sculptures appeared as "One to another" and "One to the next" in René Motro's book (Motro 2003), which was taken from Snelson's description of the sculptures, and as "Moving Column, 1st study" and "Moving Column, 2nd study" in Eleanor Heartney's essay (Heartney 2013 [2009]). The third structure was a stiff column consisting of two X-shaped modules made of plywood and monofil, where one of the modules was suspended above another in a network of 16 taut nylon threads. The discontinuity between the modules and their heavy appearance relative to the threads gave the impression that the top module was floating in the air. This column was called *Eaerly X-Piece* and was dated by Snelson to December 1948 (Gough 1998). Describing the evolution of thought in creating X-Piece, Snelson wrote from the letter from Kenneth Snelson to René Motro:

> One step leading to next, I saw that I could make the structure even more mysterious [than Moving Columns] by tying off the movement altogether, replacing the clay weights with additional tension lines to stabilise the modules one to another, which I did, making "X", kite-like modules out of plywood. Thus with forfeiting mobility, I managed to gain something even more exotic, solid elements fixed in space, one-to-another, held together only by tension members.
>
> (Snelson 1990; Motro 2003)

The remarkable features of X-Piece included discontinuity of the compression members (i.e. the X-shaped modules) and the ability to maintain its configuration irrespective of the orientation of the base (i.e. the direction of the gravity load). As a result, X-Piece embodied the two main principles of tensegrity.

During the winter of 1949, Kenneth Snelson corresponded with Fuller and sent him photographs of the three small structures (Snelson 1990; Motro 2003). Fuller suggested to Snelson to return to Black Mountain College next summer for another session. In June 1949, Snelson showed Fuller his

X-Piece. In his letter to Motro, Snelson describes Fuller's reaction to Early X-Piece as follows:

> When we got together again in June I brought with me the plywood X-Piece. When I showed him the sculpture, it was clear from his reaction that he hadn't understood it from the photos I had sent. He was quite struck with it, holding it in his hands, turning it over, studying it for a very long moment. He then asked if I might allow him to keep it. It hadn't been my intention to part with but I gave it to him, partly because I felt relieved that he wasn't angry that I had employed geometry (Buckminster Fuller's geometry) in making art. That original small sculpture disappeared from his apartment, so he told me at the end of the summer.
> (Snelson 1990; Motro 2003)

Fuller was impressed by the discontinuity of compression and continuity of tension in Snelson's X-Piece and recognized its significance for the structural representation of the tensegrity principle, which he unsuccessfully sought after for many years (Fuller 1961). The events that occurred after this point are the source of controversy between Buckminster Fuller and Kenneth Snelson lasting for more than 30 years. In December 1949, Fuller wrote to Snelson saying he mentioned Snelson in his public lectures as a creator of the prototype demonstrating structural advantages of discontinuous compression and continuous tension and that Snelson's name would be noted in history (from a letter dated 22 December 1949, cited in Motro 2003). To this, however, Fuller added that no one else but him (Fuller) could spot the potential of Snelson's creation. In January 1951, Fuller did not mention Snelson in a publication *Architectural Forum Magazine* that contained a photograph of a tetrahedral mast, which was also built by Snelson in 1949 on Fuller's request using metal curtain rods instead of plywood. In the tetrahedral mast, the kite-like modules of the X-Piece were substituted with several tetrahedral modules, in which the struts extended from a central joint to the vertices and the cables shaped the edges of the tetrahedrons. A photograph of this mast also appeared in Fuller's book *The Dymaxion World of Buckminster Fuller* (Fuller and Marks 1973). When Snelson enquired about not being mentioned, Fuller said, "Ken, old man, you can afford to remain anonymous for a while" (cited in Snelson 1990; Motro 2003). Only years late in 1959, Snelson managed to get (with the help of Fuller's assistant and the museum curator) a public acknowledgement for his contribution to the invention of the tensegrity structural system from Fuller during the exhibition of Fuller's work at the Museum of Modern Art (MOMA) in New York, which contained a 30-feet high tensegrity mast. Two years later, Fuller credited Snelson again:

> While I have been aided by many in the development of my invented implementations of my discovery of the comprehensive and coordinating

system of nature, an extraordinary intuitive assist at an important moment in my exploration of the thus discovered discontinuous-compression, continuous-tension structures was given me by a colleague, Kenneth Snelson, and must be officially mentioned in my formal recital of my "Tensegrity" discovering thoughts.

(Fuller 1961)

Fuller mostly avoided crediting Snelson in his later publications and correspondence. For example, Snelson was mentioned neither in Fuller's most celebrated book about tensegrity, *Synergetics* (Fuller 1975) nor in his correspondence with Burkhardt (Burkhardt 2011). Fuller's failure to widely acknowledge Snelson was most likely the result of Fuller's conviction that without his crucial input, Snelson would not be able to understand the importance of X-Piece and recognize in it an entirely new structural system. To emphasize that it was his invention, Fuller coined the word "tensegrity" (a combination of "tensional integrity") or as Snelson put it: "Creating this strange name was his strategy for appropriating the idea as his own" (e.g. Coplans 1967; Schneider 1977). Furthermore, Fuller's viewpoint on tensegrity can be understood in his correspondence with Snelson in 1980, where Fuller stated that tensegrity is a cosmic law of the universe and could not have been invented but discovered (Gómez-Jáuregui 2009).

The answer to the question of who invented tensegrity is not straightforward. Stephen Kurtz wrote: "If Fuller acknowledges his debt to Snelson for the invention of the tensegrity principle, Snelson likewise acknowledges his own debt to Fuller's visionary work" (Kurtz 1968). Valentín Gómez-Jáuregui added that "the synergy . . . created by both the student and professor, resulted in the origin of tensegrity" (Gómez-Jáuregui 2009). René Motro argued that "If we are in depth to Fuller for the concept, in the author's opinion the birth of its application to space structures comprising struts and cables seems to be the result of Snelson's work" (Motro 2003). Cornel Sultan thought in a similar way, saying "If Snelson invented the object, Fuller was the one to name it, creating the word tensegrity" (Sultan 2009). In other words, Snelson saw tensegrity as a structural system used in the artistic creation of physical objects, while Fuller expanded it to a universal principle that was encountered in nature and was applicable to science, engineering and architecture. The separation between the tensegrity structure and principle puts Kenneth Snelson in competition with Karl Ioganson. To resolve this matter, Maria Gough reasoned:

[B]oth Snelson and Ioganson discovered or invented, independently, the same structural principle. In any positivist account, Ioganson's temporal priority would grant him the greater originality, but the matter of invention is significantly more complex. What must be factored in is an assessment of what was made of the new principle by the inventor and his contemporaries. As is well known, the tensegrity principle was greatly developed in the postwar period by Snelson, for whom it became the basis of his life's work. . . . We have no record, however, that Ioganson referred to his tensegrity structure in any specific way nor did he, unlike Snelson and Fuller, patent the principle.

(Gough 1998)

The significance of Ioganson's invention was not grasped by Constructivists and therefore his work was nearly forgotten. In contrast, both Snelson and Fuller developed and popularized the tensegrity concept in their own ways. Using his artistic background and talent for spatial construction, Snelson created numerous tensegrity sculptures many of which were non-conventional, non-symmetrical and irregular (e.g. see Heartney 2013 [2009]). Fuller implemented the tensegrity concept in his masts, spheres and geodesic domes and promoted it through his numerous publications and inspirational lectures given to large audiences. Fuller developed the first tensegrity structure based on polyhedral shape. This was an expanded octahedron called by Fuller "six-islanded-strut icosahedron Tensegrity" (Sadao 1996). Fuller's students developed a series of tensegrity structures based on polyhedral. This included "vector equilibrium" (in Fuller's denomination) or cuboctahedron, "thirty-islanded Tensegrity sphere" or icosahedron, "six-islanded Tensegrity tetrahedron" or truncated tetrahedron and "three-islanded octa-Tensegrity" or triacontahedron (Sadao 1996; Gómez-Jáuregui 2010). In 1953, Fuller developed a 90-strut tensegrity enenicontahedron at Princeton University and a 270-strut tensegrity sphere at the University of Minnesota. The first Fuller's tensegrity sphere was constructed with the help of students in 1959 at Oregon University School of Architecture (Fuller and Marks 1973).

David Emmerich also claimed that he was the inventor of the tensegrity structures which he named "self-stressing structures" (or "structures auto-tendantes", see Emmerich 1963). David Emmerich worked in France independently of Buckminster Fuller and Kenneth Snelson. It should however be noted that Emmerich's work on tensegrity structures was probably inspired by Iognason's Equilibrium structure which he observed in the photographs of Moholy-Nagy's book (Moholy-Nagy 1929; Emmerich 1988). In his work, Emmerich aimed to bled architecture and engineering. He researched structural morphology in architecture using geometric combinatory analysis and focusing on regular partitioning in space, natural shapes and their structural stability. In the 1950s, Emmerich used morphogenesis for investigating the laws controlling the development of architectural form, which eventually led to the invention of a tensegrity structure in 1958 (Chassagnoux 2006). This invention took place in a hospital where Emmerich spent a hundred days. To pass the time, he used Mikado biscuit sticks and pieces of thread to construct initially planar models and then supported masts and eventually a self-tensioning tripod (or Simplex), a tetrapod and a pentapod. After initial experiments with simple tensegrity prisms, Emmerich moved to create complex tensegrity structures. He thought that forms are self-organizing "geometric beings in space" organized according to specific laws and that morphogenesis is the underlying principle of tensegrity. Emmerich developed several rules for assembling complex tensegrity structures from simple prismatic modules based on stacking, penetration and close or infinite braiding. The application of these assembly rules enables the construction of self-contained and propagating structures (Chassagnoux 2006). The self-contained structures consisted of polyhedrons and hyperpolyhedrons. The propagating structures could develop (i) linearly resulting in masts and trusses, (ii) bidirectionally producing double-layer lattices, or (iii) spatially yielding complex lattices composed of polyhedrons and hyperpolyhedrons connected by masts. Emmerich popularized the tensegrity structures through seminars, publications and conference lectures (e.g. Emmerich 1967a, 1967b) and by creating regular, symmetric tensegrity structures that also included domes.

One way of establishing the authorship of the tensegrity system is to consider the chronology of registration and gaining the patents. Despite being the first to build a tensegrity structure in 1921, Karl Ioganson did not receive or register any patent and so unfairly he is omitted. Both Emmerich and Fuller registered their first patents in 1959 (Emmerich 1959; Fuller 1962). Emmerich's patent was not registered correctly and subsequently was not granted, while Fuller's patent was granted in 1962. Emmerich applied for a second patent in 1963, which was granted in 1964 (Emmerich 1963). Snelson was the last to register his patent in 1960 and was granted in 1965 (Snelson 1965). It is important to note at this point that Fuller, Snelson and Emmerich developed and patented the same basic tensegrity system consisting of three struts and nine cables, that is, the Simplex.

Definition of tensegrity

There is no consensus in the definition of tensegrity. The differences in the approaches to the tensegrity structural system were, for example, reflected in many names given to the simplest tensegrity structure shown in Figure I.1, which includes Simplex, Elementary Equilibrium, 3 struts T-prism, "3 struts, 9 tendons", 3 strut single-layer, twist element, twist unit, regular triplex, etc. (Motro 2003; Gómez-Jáuregui 2010). Using different approaches, researchers developed many definitions that emphasize different aspects of tensegrity. It should be noted that most definitions described tensegrity structural system, although several were given to the tensegrity principle.

Throughout his career, Fuller suggested many definitions of tensegrity. He described the tensegrity principle as follows:

> All structures properly understood, from the solar system to the atom, are tensegrity structures.

> (Fuller 1975)

Islands of compression inside an ocean of tension.

(Fuller 1955, 1962)

An assemblage of tension and compression components arranged in a discontinuous compression system.

(Fuller 1975)

Fuller also proposed many rather intuitive definitions of the structural system, for example:

Tensegrity is an inherently nonredundant confluence of optimum structural-effort effectiveness factors. Tensegrity structures are pure pneumatic structures, at the subvisible level of energy events.

(Fuller 1975)

Tensegrity describes a structural-relationship principle in which structural shape is guaranteed by the finitely closed, comprehensively continuous, tensional behaviors of the system and not by the discontinuous and exclusively local compressional member behaviours.

(Fuller 1975)

Tensegrity . . . [is] a structure the shape of which is guaranteed by the tensional behaviour of the system, and not by the compressional behaviors.

(Fuller, cited in Applewhite 1986)

A more technical definition can be found in his patent:

[A] plurality of discontinuous compression columns arranged in groups of three non-conjunctive columns connected by tension elements forming tension triangles.

(Fuller 1962)

Emmerich's definition of the tensegrity structural system given in his patent was very technical:

Self-stressing structures consist of bars and cables assembled in such a way that the bars remain isolated in a continuum of cables. All these elements must be spaced rigidly and at the same time interlocked by the pre-stressing resulting from the internal stressing of cables without the need for external bearings and anchorage. The whole is maintained firmly like a self-supporting structure, whence the term self-stressing.

(Emmerich 1963)

Snelson defined the tensegrity structural system in his patent in a more descriptive manner:

> The present invention relates to structural frameworks and more particularly to a novel and improved structure of elongated members with a care separately placed, either in tension or in compression to form a lattice, the compression members being separated from each other and the tension members being interconnected to form a continuous tension network.
>
> (Snelson 1965)

Later, he rephrased his definition and included more structural details:

> Tensegrity describes a closed structural system composed of a set of three or more elongate compression struts within a network of tension tendons, the combined parts mutually supportive in such a way that the struts do not touch one another, but press outwardly against nodal points in the tension network to form a firm, triangulated, prestressed, tension and compression unit.
>
> (Snelson 1996)

To this, Snelson added: "Tensegrity structures are endoskeletal prestressed structures—and that restriction leaves out endless numbers of items" (Gómez-Jáuregui 2010).

Whiteley and Connelly and their colleagues took a more mathematical approach:

> A tensegrity framework consists of bars which preserve the distance between certain pairs of vertices, cables which provide an upper bound for the distance between some other pairs of vertices and struts which give a lower bound for the distance between still other pairs of vertices.
>
> (Roth and Whiteley 1981)

> A tensegrity framework is an ordered finite collection of points in Euclidean space, called a configuration, with certain pairs of these points, called cables, constrained not to get further apart; certain pairs, called struts, constrained not to get closer together; and certain pairs, called bars, constrained to stay the same distance apart.
>
> (Connelly and Whiteley 1996)

> Tensegrities have a purity and simplicity that lead very naturally to a mathematical description. Putting aside the physical details of the construction, every tensegrity can be modelled mathematically as a configuration of points and vertices, satisfying simple distance constraints.

Snelson's structures are held together with two types of design elements (engineers may say members), which can be called cables and struts. Two elements play complementary roles: Cables keep vertices close together: struts hold them apart.

(Connelly and Back 1998)

Pugh suggested a concise, mechanically accurate and perhaps the most widely accepted definition:

A tensegrity system is established when a set of discontinuous compressive components interacts with a set of continuous tensile components to define a stable volume in space.

(Pugh 1976)

Many more definitions of the tensegrity structural system reflecting different perspectives of the researchers can be found in scientific and engineering literature. Several researchers described the tensegrity structural systems as prestressed, pin-joined cable networks:

Internally prestressed, free-standing pin-joined networks, in which the cables or tendons are tensioned against a system of bars or struts.

(Hanaor 1987)

A tensegrity structure is any structure realised from cables and struts, to which a state of prestress is imposed that imparts tension to all cables . . . as well as imparting tension to all cables, the state of prestress serves the purpose of stabilizing the structure, thus providing first-order stiffness to its infinitesimal mechanisms.

(Miura Koryo and Sergio Pellegrino, cited in Tibert 2002)

These descriptions were expanded to include the continuity of tension members and the discontinuity of compression members, where the network of tension members enveloped the compression members:

A tensegrity system is a system in a stable selfequilibrated state comprising a discontinuous set of compressed components inside a continuum of tensioned components.

(Motro 2003)

Tensegrity systems are free-standing pin-jointed cable networks in which a connected system of cables are stressed against a disconnected system of struts and extensively, any free-standing pin-jointed cable networks composed of building units that satisfy aforesaid definition.

(Wang and Li 1998, 2003)

Tensegrity is a structural principle based on the use of isolated components in compression inside a net of continuous tension, in such a
way that the compressed members (usually bars or struts) do not touch
each other and the prestressed tensioned members (usually cables or
tendons) delineate the system spatially.

(Gómez-Jáuregui 2010)

Some researchers emphasized that the tensegrity structural systems should
be in a state of self-stressed equilibrium:

Tensegrity systems are systems whose rigidity is the result of a state of
self-stressed equilibrium between cables under tension and compression elements.

(Motro 1992)

Selfstressed equilibrium cable networks in which a continuous system of cables (tendons) are stressed against a discontinuous system of
struts.

(Wang 1998)

Tensegrity Systems are spatial reticulate systems in a state of self-stress.
All their elements have a straight middle fibre and are of equivalent
size. Tensioned elements have no rigidity in compression and constitute a continuous set. Compressed elements constitute a discontinuous
set. Each node receives one and only one compressed element.

(Motro 2003)

While others highlighted the importance of the configuration of members
within the tensegrity structures:

A tensegrity system is composed of any given set of strings connected
to a tensegrity configuration of rigid bodies.

(Skelton and de Oliveira 2009)

A tensegrity is a pattern integrity which has purely-tensile portions which are essential to its integrity. A "pattern integrity" is a
description of a system whose instances maintain a stable pattern
in space and time in a variety of situations. 'Purely-tensile portions'
means portions which are at least sometimes in tension and are
not required to sustain non-tensile loads. 'Essential to its integrity'
means that in general the instances of pattern integrity are not able
to maintain their stable patterns without the presence of certain
purely-tensile portions.

(Burkhardt 2011)

[Tensegrity] structures in which form and function truly are . . . one, and the visible configuration of the sculpture is simply the revelation of otherwise invisible forces.

(Heartney 2013 [2009])

As a result, the definition of tensegrity varies from a fundamental principle of nature to define a very specific type of structures. The broad definition of tensegrity proposed by Fuller can be applied to the universe, to the planetary system or to the structure of the atom. On the other hand, tensegrity can also be applied to define a very specific type of structures. It is proposed here to distinguish between two terms of tensegrity: the so-called principle of tensegrity and the so-called tensegrity structures. The principle of tensegrity was already defined few times by Fuller and can be used by philosophers and scientists to study the planetary system, the movement of the electron about the atom nucleus or the expansion of the universe and the pulling of the galaxies apart. It is proposed here to use the term tensegrity structures to a very limited family of structures (Snelson type structures "floating compression") and to exclude all structures with extensive cables, such as cable nets, suspension bridges, etc. The definition is as follows:

Tensegrity structures are composed of a continuous tensile net (cables) and compression elements (bars) which are connected to the tensile net only and not to each other. They are constructed with no or only very few compression elements attached to the foundations. They are prestressable: prestressing induces tension to the cables and compression to the bars.

Main features of tensegrity structures

The main features of the tensegrity structural system can be summarized as follows:

1. *Pin-jointed connections:* All members are connected at the nodes by pin-joints (or frictionless ball-joints not capable of transferring moments and torque). The members do not have any contact with each other along their length.
2. *Loaded connections:* All loads and supporting conditions are applied at the nodes.
3. *Singular structural function:* Each member can transfer only compression or tension force along a line passing through its nodes. Tension members cannot become compression members.
4. *Member shape:* In general, tension members can be in the shape of a straight line (i.e. tension ties) or a surface (i.e. tension membranes in canopies supported by masts), while compression members are in

the shape of a straight line (i.e. compression struts), a curved line (i.e. beam-columns; e.g. Schorr et al. 2021), a surface (i.e. plates and shells; e.g. Ma et al. 2022) or a volume (i.e. a volume of compressed air in pneumatic structures). In certain cases, discussed later in the classification of tensegrity structures, clusters of struts connected into planar closed polygonal modules can be considered as single complex compression members. It should be noted that the member shape is less important than its structural function. The classical tensegrity structures are composed of straight-line members as this shape is the most efficient in transferring tension and compression forces.

5. *Continuous tension and discontinuous compression (or compression localization):* The tension members are organized in a continuous network connected at the nodes and extending through the entire structure, while the compression members represent a discontinuous set (i.e. separate members or member casters). This feature was figuratively described by Fuller as "islands of compression inside an ocean of tension" (Fuller 1955, 1962).

6. *State of self-stress:* Stability is achieved through prestressing the members during construction, which results in every member being under either tension or compression force according to its structural function. The prestress controls the stiffness, load-carrying capacity and volume a tensegrity structure occupies (Pugh 1976; Tibert and Pellegrino 2003).

7. *Self-equilibrated stability:* When loaded, the tensegrity structure is in a state of equilibrium between tension and compression forces in the members.

It is important to note that features 1, 2 and partially 4 categorize the tensegrity structures as trusses while the rest of the features listed earlier separate them into a special type of three-dimensional trusses. Further discussion of the features of the tensegrity structural system can be found in the existing engineering literature (e.g. Motro 2003; Gómez-Jáuregui 2010).

Classification of tensegrity structures

The first classification of the tensegrity structures can be attributed to Fuller, who broadly divided them into prestressed and geodetic tensegrities. The prestressed tensegrity relied on the prestress of its tension members for maintaining self-equilibrium, while the geodetic tensegrities required their structural members to be oriented along geodetic lines, that is, minimal spherical paths (Ingber 2003; Gómez-Jáuregui 2010). Pugh listed tensegrity structures of polygonal and polyhedral shapes (Pugh 1976). The structures were characterized using the positions of struts relative to the centre of the shape, the presence of cables on the faces of the shape, the grouping of struts into clusters and the number of layers in the shape. Grip suggested classifying the tensegrity structures using their shape-based geometric similarity to convex polyhedra, where the edges of the polyhedral shape were defined by the compression members of the structure (Grip 1992). The researcher also distinguished between structural configurations corresponding to single convex polyhedra and their three-dimensional arrays. Connelly and Back (1998) applied rotation and reflection to regular polyhedral shapes for producing a catalogue containing six groups of symmetric super-stable tensegrity structures (Connelly and Back 1998). Other researchers used different criteria for developing more detailed classification systems of tensegrity structures. The criteria included tensegrity cell pattern, arrangements of compression and tension members and jointing methods.

Classification based on tensegrity cell pattern

Pugh distinguishes three families of tensegrity structures using basic patterns of arrangement of their compression and tension members, which included Diamond, Circuit and Zigzag pattern.

The Diamond pattern was used for constructing single-layer prismatic and multiple-layer cylindrical structures. The name of this pattern came from the shape of a cell repeating along the side of the structure. This cell has a diamond (or rhombic) shape composed of four cables at the edges and a strut positioned along the longer diagonal. This pattern was later called "rhombic configuration" by René Motro (2003). The simplest prismatic structure of this kind is the Simplex (Figure I.1), while the expanded octahedron can be considered the simplest double-layer structure. The stability of each node of a diamond tensegrity structure requires a minimum of three cables and one strut. In his description, Pugh focused on two-, three- and four-layered cylindrical structures with up to six struts in each layer. Increasing the number of struts in the layers beyond six resulted in decreased stability of these structures. To keep stable the structures with more than four layers, Pugh suggested limiting the number of struts in each layer to three or four. Furthermore, the cables at the top and bottom faces of the multilayered structures had to be shortened resulting in barrel-shaped cylinders. The Diamond pattern was used as a base for generating the Circuit and Zigzag patterns, the implementation of which resulted in spherical tensegrity structures. The Circuit pattern was obtained through systematic cell degeneration in diamond tensegrity structures by folding certain diamond cells along the diagonal struts and removing one of the paired cables in the fold. This resulted in the ends of struts adjacent to the fold jointed together creating strut circuits, which gave the name to this pattern. The cell degeneration process can produce either multiple intertwining polygonal circuits (triangular, square, pentagonal, for example, see Photographs 4, 6 and 7 in Pugh 1976) or a single tangled multicomponent circuit (e.g. see Photograph 5 in Pugh 1976).

It is noteworthy that the presence of intertwining strut circuits prompted Fuller to name these structures as "basketry tensegrity" (Pugh 1976; Motro 2003). The strut circuits in these structures are inscribed in cable circuits of polygonal shapes with the doubled number of sides. The cables remain only on the external envelope of the structure and define the edges of a regular or semi-regular polyhedral shape. Since all tension members are on the external envelope and all the compression members are inside the structure, these are pneumatic structures, and their shape is dictated by minimum volume. It should be noted that struts combined into a polygonal circuit can be considered as a single complex compression member. After joining strut ends and removing unnecessary cables, each node in the strut circuits is stabilized by four cables connecting it to other strut circuits.

The Zigzag (or Z-shape) pattern was obtained by adding a cable along the short diagonal in the Diamond pattern cell and removing one opposite pair of cables. Since the Diamond cell has two pairs of opposite cables, two Zigzag pattern cells can be produced. These cells are mirror images of each other (i.e. the Zigzag pattern structures are enantiomorphic) and therefore can be called as Z-shape and reversed Z-shape cells. A Zigzag tensegrity structure can contain only one type of Z-shape cell. Mixing of mirror-image cells results in the loss of structural form and stability. The application of the Zigzag pattern to Diamond tensegrity structures results in their transformation into different polyhedra. For example, the Diamond expanded octahedron is transformed into a Zigzag truncated tetrahedron. It is important to note that the application of the Z-shape or reversed Z-shape transformation to Diamond tensegrity structures with faces of different polygonal shapes results in Zigzag structures of different polyhedral shapes because it expands the different sets of polygonal faces in the original structure. For example, a cuboctahedron has triangular and square faces. Depending on the type of Z-transformation, it can become either a truncated cube (after expanding its square faces) or a truncated octahedron (after expanding its triangular faces).

The Zigzag transformation is often accompanied by the loss of regularity or distortion of some cable-formed polygonal faces because the struts approach the vertices of these faces from different angles (usually two angles). As a result, the struts apply non-uniform constraints on the external network of cables forcing it, like a balloon skin, into the smallest volume containing distorted faces. Pugh reported that pentagonal faces often become irregular, like in the case of the truncated tetrahedron, while the square, hexagonal, octagonal and decagonal faces in the structures based on Archimedean polyhedra always distort. The distortion can be eliminated by introducing additional cables between the ends of struts approaching the face vertices at similar angles. Pugh also noted that the structures with distorted faces were statically stable despite having only two-thirds of the number of members required for the stability of other pin-joined structures such as frames (see Maxwell's rule in Maxwell 1864; Calladine 1978). Furthermore, their dynamic stiffness was low as they readily vibrated when cables were plucked. The excessive vibrations disappeared after adding cables to the distorted faces which usually increased the number of members to the minimum required for frame stability. Pugh also noted that the ability to readily vibrate indicates an easy transfer of loads throughout these structures without any load localization.

Both the Circuit and Zigzag patterns can be used for producing geodetic polyhedra through triangular discretization of faces of existing polyhedra (e.g. Platonic polyhedra) and the introduction of additional struts and cables. The discretization begins with the division of all polygonal faces into triangles followed by their subdivision into smaller triangles. This process was governed by the frequency that was formulated as the number of segments which each edge of the original face was divided into. The Circuit pattern geodetic tensegrity structures were generated by the discretization of faces with a frequency multiple of two. The geodetic strut circuits took upon multifaceted polygonal shapes. The Zigzag pattern geodetic tensegrity structures required the discretization frequency multiple of three.

The degeneration of cells in the Circuit pattern transformation results in the compression of an original Diamond pattern structure. Pugh mentioned that the application of this transformation to an expanded octahedron resulted in its compression into a regular octahedron (see Figure 10 in Pugh 1976). In the generated structure, the six struts will be rearranged into three pairs, which can be seen as two-sided polygonal circuits. After removing one unnecessary strut from each pair and introducing a joint in the central intersection of struts, the Circular pattern tensegrity octahedron becomes Ioganson's octahedron, that is, structure III in Gough's numbering of Ioganson's spatial constructions (Gough 1998). Furthermore, the application of the Zigzag pattern to the expanded octahedron during the Circular pattern transformation without introducing the central joint results in Ioganson's spatial construction (structures II in Gough's numbering, see Gough 1998, and see also Diagram 6.2 in Pugh 1976). As a result, Ioganson's spatial constructions are closely associated with the tensegrity structures.

The Diamond and Zigzag pattern tensegrity structures are based on the Diamond and Zigzag cells that consist of one strut, several cables and four nodes. Pugh suggested that new tensegrity cells can be generated by introducing additional nodes and cables (see Diagram 7.2 in Pugh 1976). Li et al. (2010) listed all possible cells consisting of one strut, four nodes and two to five cables. Researchers introduced an X-diamond (or X-rhombic) cell by adding a fifth cable along the shorter diagonal of the original Diamond cell. The X-diamond cell was used as a basis for generating nine simpler cells by removing cables. This included two four-cable cells (one of which was the Diamond cell), four three-cable cells (one of which was the Zigzag cell) and three two-cable cells. It was noted that the two-cable cells could only be used for two-dimensional tensegrity structures. In Diamond tensegrity structures, cells usually share cables. If the shared cables are considered belonging only to one cell, the Diamond cells are transformed into three-cable cells.

Classification based on compression members

Several researchers proposed to classify the tensegrity structures depending on the arrangement of compression members. This classification is based on the maximum number of compression members connected at one joint and it is described in several research works as follows:

> A tensegrity system is a stable connection of axially-loaded members. A Class k tensegrity structure is one in which at most k compressive members are connected to any node. E.g., a traditional tensegrity structure is a class 1 structure because only one compression member makes a node.
>
> (Kanchanasaratool and Williamson 2002)

> A tensegrity configuration that has no contacts between its rigid bodies is a class 1 tensegrity system, and a tensegrity system with as many as k rigid bodies in contact is a class k tensegrity system.
>
> (Skelton and de Oliveira 2009)

> A tensegrity structure can be classified based on the maximum number of strut(s) that joined at a node. By definition, the Class I tensegrity is the original concept in creating a tensegrity. Class II and III tensegrity have more applications in a situation where more stable structures are required.
>
> (Gan 2020)

Based on this classification, the Diamond and Zigzag pattern structures are Class I tensegrities, while the Circuit pattern structures are Class II tensegrities, although they can also form higher class tensegrity systems.

Classification based on joining method

Motro distinguished between two topological types of structural tensegrity systems: Spherical and Star module tensegrities (Motro 2003). The Spherical module was defined by being homeomorphic to a sphere, that is, when all cables can be mapped without intersections on a sphere circumscribing the structure. The Star module allowed for mapped cable intersections. This implies that all cables in a Spherical module tensegrity are on the external envelope of the structure. Based on this description, the Diamond, Circuit and Zigzag pattern tensegrity structures belong to the Spherical module tensegrity. The Diamond pattern tensegrity structures can be classified as Prismatic module tensegrities, which is a subset of the Spherical module tensegrity.

The tensegrity modules can be joined into complex structures. Motro suggested three ways for joining the Spherical modules, which result in one-, two- and three-directional tensegrity structures (Motro 2003). These descriptions define the number of axes governing structural geometry. The one-directional structures can be linear or curved and include masts, towers, arches and tori. The notable early examples of the one-directional structures include Fuller's tensegrity masts (e.g. North Carolina State College Tensegrity Mast built in 1950, University of Oregon Tensegrity Mast built in 1953, see Motro 2003), many Snelson's productions (e.g. X-Piece, X-Column, Needle Tower, Tower of Light, Cantilever, see Heartney 2013 [2009]), Pugh's cuboctahedral mast, Boulez's "Chinese mast", Chassagnoux's "Mat Autotendant", Motro's arch and torus (Pugh 1976; Motro 2003), while more recent examples include many Snelson's sculptures (e.g. Rainbow Arch, Penta Tower, Black E.C. Tower, Sleeping Dragon, see Heartney 2013 [2009]), other Snelson's sculptures (e.g. Rain b Tensegrity Tower in Rostock (Schlaich 2004), von Richthofen's Dubai Tensegrity Tower (AurelVR 2008) and Santiago Antenna Tower (ArchDaily 2014a). It is interesting to note that the Needle tower was a Diamond pattern structure made of three-strut modules, while the "Chinese mast" was made of Simplex modules.

The two-directional tensegrity structures can be divided into planar, single-curvature and double-curvature systems. Pugh proposed plane two-directional structures built of tetrahedral, octahedral and icosahedral tensegrity modules (see Diagrams 8.1, 8.3 and 8.4 in Pugh 1976) and built prototypes of tetrahedral and octahedral module structures (see Photographs 27 and 28 in Pugh 1976). It should be noted that these proposals contained gaps in the modular grid. Fuller's geodetic domes and spheres are double curvature single-layer tensegrity grids. Vilnay developed single-layer tensegrity grids that could cover large spans without struts touching each other within tensegrity modules (Vilnay 1990). The type of the two-directional tensegrity structures that attracted the most attention has been the plane and curved double-layer tensegrity grids that were typically composed using three- or four-strut modules. However, the uses of the five- and six-strut modules were also explored (Hanaor 1991a, 1991b). Many researchers developed their own proposals, built prototypes and investigated their structural behaviour. Emmerich was the first to propose a two-directional tensegrity grid in his first patent (Emmerich 1963) but did not provide any structural analysis showing that the grid was stable (Motro 2003). Emmerich also provided geometrical descriptions of "self-stressed planar nets" in his book (Emmerich 1988). The notable early examples of plane double-layer tensegrity grids include those developed by Snelson (e.g. Triangle and Square Planar pieces as well as Planar Wave piece exhibited in 1960, see Snelson 2012) as well as Motro, Hanaor and Kono (Motro et al. 1986b; Motro 1987; Hanaor 1991a, 1991b, 1992, 1993; Kono et al. 2000). The Kono's proposal consisted of three-strut (distorted Simplex) modules, four-strut (half-cuboctahedral or prismatic) modules were used in the Motro's and Hanaor's prototypes, while Snelson experiment with both three- and four-strut modules. The recent developments in the field of plane double-layer grids include, for example, the works of Wang and colleagues (e.g. Wang and Li 2003; Wang 2004), Gómez-Jáuregui and colleagues (e.g. Gómez-Jáuregui 2010; Gómez-Jáuregui et al. 2012, 2013; Quilligan et al. 2020) and many others (e.g. Obara 2019; Obara and Tomasik 2020, 2021).

The single-curvature double-layer grids were first proposed by Motro in the 1980s (Motro 1990). The researcher developed a technique for curving planar double-layer grids into a cylindrical shape through altering the arrangement of struts in the comprising modules while keeping the modules in a self-stressed state. Motro reported that four-strut modules could be rearranged in two different ways leading to either convex or concave shapes. In the 1980s, Motro also proposed a method for developing a double positive curvature double-layer grid by mapping the horizontal cables of a planar double-layer grid on two parallel double positive curvature surfaces using Formex algebra (Motro 2003). The struts and inclined bracing cables were added later to the dome-shaped system. Motro built a prototype of a double curvature double-layer grid, where the curvatures of the grid could be changed by altering the lengths of cables (see Figure 4.57 in Motro 2003). Hanaor also developed proposals for single and double positive curvature double-layer grids composed of three-strut modules (Hanaor 1992, 1993, 1994). Olejnikova investigated single- and double-curvature double-layer grids composed of three- and four-strut modules by mapping them on different surfaces such as cylinders and spheres (Olejnikova 2012). In addition, researcher also proposed saddle-shaped double-layer grids mapped on surfaces with negative Gaussian curvatures.

The three-directional tensegrity structures were described by Pugh as conglomerations of modules similar to a cluster of soap bubbles (Pugh 1976). Pugh proposed a three-directional tensegrity structure built of four-strut tensegrity cubes (see Diagram 8.2 in Pugh 1976). However, these structures are the hardest to build due to developing distortions and innate enantiomorphism and therefore exist only in proposals.

The construction of multi-modular tensegrity structures required developing method for joining the modules. Pugh proposed an approach for joining tensegrity modules based on their polyhedral geometry. For example, tetrahedral tensegrity modules were joined using the vertex-to-vertex approach, while the cubic, octahedral and icosahedral tensegrity modules were joined using the edge-to-edge (or rather face-to-face) approach (see Diagrams 8.1– 8.4 in Pugh 1976). In both cases, the connections between modules required additional bracing with a set of cables arranged in triangles. Several methods were later developed for joining the tensegrity modules based on module components, that is, nodes (joints) and cables. These included the node-on-node, node-on-cable and cable-on-cable (partial and total) methods (Motro 2003; Li et al. 2010). Hanaor was one of the first to describe these methods (Hanaor 1987, 1990, 1992). Since each tensegrity module is self-equilibrated, the resulting structure is also in the state of self-equilibrium. A stable junction between modules could require an additional bracing cable as in the case of the node-on-cables jointing methods (Hanaor 1992; Motro 2003).

Classification based on tension members

The one-directional tensegrity structures composed of similar modules can be divided according to the arrangement of tension members. To implement this subdivision, three types of tension members can be introduced (e.g. see Sultan et al. 2001, 2002a, 2002b). The tension members connecting the ends of compression members from the adjacent sides of two modules (i.e. the top side of one module and the bottom side of the next module) can be called "saddle" cables and can also be identified with the letter "S". The tension members connecting the ends of compression members at similar sides of two modules (i.e. the top side of one module and the top side of the next module) can be called "vertical" cables or "V" in short. The tension members connecting the ends of compression members on the opposite sides of one module can be called "diagonal" cables or "D" in short. Using this identification system, the one-directional tensegrity structures with modules connected by all three types of tension members can be classified as SVD systems. In certain configurations, the vertical cables coincide with the diagonal cables or are missing altogether which results in the SD tensegrity systems. In tensegrity structures with modules containing connected compression members, only the SS structural type is possible.

A separate class of tensegrity structures is characterized by the network of tensile cables being partially or entirely substituted by tensile membranes. As a result, this class is commonly called membrane tensegrity structures. The simplest members of this class are a kite and an umbrella. The most obvious way of using the membrane is to place it in the external tensile envelope of the structure. The early use of membranes in tensegrity structures can be traced to Pugh who implemented it in a geodetic dome (see Photograph 31 in Pugh 1976). Nowadays, membrane tensegrity structures are often used by artists due to high visual impact (e.g. Underwood Pavilion, MOOM pavilion, see ArchDaily 2014b; Riether and Wit 2016; Feng 2019). An example of a structure with cables only partially substituted by the membranes is von Richthofen's Dubai Tensegrity Tower (AurelVR 2008). The principles of membrane tensegrity have also been widely used in fabric canopies. A subclass of membrane tensegrity structures is pneumatic structures in which the compression members are substituted with air. These structures are beyond the scope of this book.

Development of tensegrity structures

The works and popularization of tensegrity by Fuller, Snelson and Emmerich had a profound effect on the architectural, engineering and scientific communities. For example, Snelson's Needle Tower exhibited in the Hisshom Museum in Washington sparked interest in Motro (Motro 2003), which led to Motro becoming one of the most important experts in tensegrity structures. Motro published many scientific works on the subject and carried out many research projects (e.g. Motro et al. 1986a, 1986b; Motro 1987, 1990, 1992, 2003). The list of Fuller-inspired influential researchers includes Pugh and Kenner. Pugh chose a practical approach by showing methods for constructing prototypes of known tensegrity structures and developing new ones (Pugh 1976). He also provided detailed descriptions of various existing tensegrity structures and developed pattern-based classification. In contrast, Kenner explored tensegrity structures using an analytical approach based on geometry, mathematics, and engineering mechanics (Kenner 1976). Kenner treated these structures as spatial diagrams of forces whose equilibrium defined stable structural configurations. For example, Kenner found a stable prestressable configuration for the expanded octahedron using node equilibrium conditions and symmetry.

The static behaviour of tensegrity structures was investigated in the 1970s, 1980s and 1990s very extensively. Fuller, Emmerich and Pugh approached the tensegrity structures using their experience, intuition and geometrical considerations without considering their structural equilibrium (Fuller 1975; Emmerich 1988; Pugh 1976). Calladine was the first to describe the tensegrity structures as three-dimensional pin-jointed trusses and investigate their statics behaviour by using matrix analysis (Calladine 1978). Calladine's approach was used by many other researchers who applied matrix-based methods for studying statics of tensegrity structures, which included discovering and form-finding new tensegrity configurations (e.g. Kenner 1976; Tarnai 1980; Hanaor 1988), and studying first- and higher-order infinitesimal and finite mechanisms, equilibrium configurations, self-stress states and resolvable and non-resolvable forces (e.g. Pellegrino and Calladine 1986; Pellegrino 1989; Vilnay 1990; Calladine and Pellegrino 1991, 1992). Matrix analysis was also used to formulate the dynamic response of tensegrity structures (Vilnay 1990). Other researchers selected a mathematical approach, describing the tensegrity structures as constrained clusters of points in the Euclidian space (e.g. Connelly 1980; Roth and Whiteley 1981; Connelly 1982; Connelly and Whiteley 1996; Connelly and Back 1998). These works introduced several new concepts, including first- and second-order rigidity, prestress stability and super-stability as well as used the well-known concepts of infinitesimal and static rigidity. Connelly and Whiteley (1996) established hierarchical relationship between different order rigidities of tensegrity structures. They stated that to achieve rigidity, a tensegrity structure should initially have first-order (i.e. infinitesimal or statics) rigidity, then fulfil prestress stability and finally should be second-order rigid. It should be noted that later Juan and Tur (2008) extended this hierarchical sequence by introducing stability and material criteria, including super-stability, universal global stability, global stability and unyielding between the conditions of first-order rigidity and prestress stability and adding third-order rigidity as the final condition. Roth and Whiteley (1981) applied the graph theory for developing new tensegrity structures. Connelly (1982) used an energy approach which led to redefining the rigidity concept based on the positive definiteness of the stress matrix. The energy approach was also applied by Salerno (1992) for recognizing infinitesimal mechanisms and by Connelly and Whiteley (1996) for linking material properties, self-stress configuration and external loads applied on a tensegrity structure. Grip developed new super-stable polyhedral tensegrity structures using symmetric dual transformation which was based on expanding one set of polygonal faces and contracting another set (Grip 1992). The investigation of the dynamic behaviour of tensegrity structures in the 1970s and 1980s was limited to only a few works. Pugh was the first to point out that tensegrity structures can easily transfer vibration (Pugh 1976). Motro and colleagues tested the non-linear static and dynamic behaviour of the Simplex (Motro et al. 1986a). Since 1990s, the research in tensegrity structures underwent impressive development and diversification due to the significant increase of computational capabilities in terms of both more powerful hardware and software. This allowed for the development of more sophisticated modelling algorithms cable of considering the material and geometric non-linearity in statics and dynamics as well as for carrying out the numerical analysis more efficiently. The technological advances led to discovering mechanisms in the prototypes of tensegrity structures (e.g. Sultan and Skelton 1998a, 2004). The listed developments allowed researchers to investigate non-linear static (e.g. not capable of transferring moments and torque). The members do not have any contact with each other along their lengths. Further information on the developments in the field of tensegrity structures can be found in Sultan (2009), Gómez-Jáuregui (2010) and Micheletti and Podio-Guidugli (2022).

An important task in understanding the static behaviour of tensegrity structures is to find their shape. Kenner, Tarnai and Hanaor developed analytical solutions for form-finding of tensegrity structures with simple symmetrical geometries (Kenner 1976; Tarnai 1980; Hanaor 1988). Sultana and colleagues generalized the form-finding problem using innate characteristics of the tensegrity structures and attempted to obtain closed-form solutions of the associated prestressability problems using symbolic mathematical modelling (Sultan et al. 2001; Sultan and Skelton 2003b). Motro was the first to apply the dynamical relaxation (i.e. numerical) method for form-finding prestressable configurations of tensegrity prisms (Motro 1984; Motro et al.

1986a). This method was reliably applied for form-finding tensile structures by Barnes (1999) and was refined by Zhang et al. (2006) and applied for form-finding irregular tensegrity structures. Later, powerful numerical techniques, including the force–density method and its modifications, were introduced (e.g. Schek 1974; Vassart and Motro 1999; Masic et al. 2005; Zhang and Ohsaki 2006; Estrada et al. 2006; Tran and Lee 2010, 2013; Lee and Lee 2014). Nishimura and Murakami (2001) developed an analytical method for form-finding configurations of cyclic frustum tensegrity modules with an arbitrary number of stages using their symmetric property. Micheletti and Williams (2007) developed a marching numerical algorithm for solving the form-finding problem of tensegrity structures, which included foldable, deployable and variable-geometry structures. Rieffel et al. (2009) applied an evolution algorithm for form-finding large tensegrity structures. Xu and Luo (2010) considered the form-finding problem as an optimization problem with geometrical and material constraints. The researchers used this approach for form-finding irregular tensegrity structures. Pagitz and Tur (2009) developed a form-finding algorithm that was based on the finite element method and considered material stiffness, geometry and cable prestress. The form-finding process minimized the potential energy of the system and preserved the total cable length. Xu et al. (2016) used mixed integer linear-quadratic programming to solve a topology optimization problem associated with form-finding of tensegrity structures. Problem solution accounted for the self-stressed state, node stability and geometric constraints on member distribution and symmetric configuration. Ma et al. (2022) developed a form-finding procedure based on solving the non-linear equilibrium equations using a modified tangent stiffness matrix and line search algorithm. More detailed information on the form-finding methods can be found in the review works of Tibert and Pellegrino (2003), Juan and Tur (2008) and Micheletti and Podio-Guidugli (2022).

Tensegrity structures saw first applications as sculptures (e.g. by Ionanson and Snelson) and structural prototypes demonstrating the tensegrity principles (e.g. structures created by Fuller and Emmerich). Due to their advantages and unique nature, these structures offer potential solutions for many practical problems. Nowadays, the applications of tensegrity structures are rapidly expanding into different fields of engineering and science, which cover structural engineering, deployable structures, robotics, energy absorption, metamaterials, biology, etc. Examples of structural engineering applications of tensegrity principles include the Olympic Fencing and Gymnastics Arena in Seoul (completed in 1988, TensiNet 2022), Georgia Dome in Atlanta (completed in 1992, BIRDAIR 2022), Tensegrity Tower in Rostock (completed in 2003, Schlaich 2004), Kurilpa Bridge in Brisbane (completed in 2009, ARUP 2022), La Plata Stadium in Buenos Aires (completed in 2011, Stadium Guide 2022), Dubai Tensegrity Tower completed in 2008 (AurelVR 2008), Santiago Antenna Tower (completed in 2017, ArchDaily 2014a), etc.

The ideas of folding and deploying tensegrity structures were pioneered by Furuya (1992) and Hanaor (1993) and developed by Motro and colleagues (e.g. Bouderbala and Motro 1998; Smaili et al. 2004) and Sultan and Skelton (e.g. Sultan and Skelton 1998b, 2003a). Lightweight and high resilience combined with foldability and deployability made the tensegrity structures extremely attractive for the space industry (e.g. Tibert and Pellegrino 2002; Tibert 2002; Stern 2003; Zolesi et al. 2012; Scolamiero et al. 2015, 2017; Ganga et al. 2016; Zawadzki and Al Sabouni-Zawadzka 2020). Examples of proposed space applications of the tensegrity structures included space stations, space telescopes and deployable antennas. Gómez-Jáuregui et al. (2018) developed a new type of deployable double-layer tensegrity grid. Hrazmi et al. (2021) applied a deployable double-layer tensegrity grid as a supporting structure in developing a platform for accessing seashore.

The potential for locomotion that derives from the ability to move struts within a tensegrity structure by changing the lengths (prestress) of cables has been exploited in robotics since the early 2000s (e.g. Aldrich et al. 2003; Paul et al. 2005; Rovira and Mirats Tur 2009; Shibata and Hirai 2009; Korkmaz et al. 2011, 2012; Bruce et al. 2014; Kim et al. 2017). Tensegrity-based systems were incorporated in rovers (e.g. Hall 2015; Sabelhaus et al. 2015), bio-mimicking robots (e.g. Lessard et al. 2016; Zappetti et al. 2020; Liu et al. 2022b) and modular soft robots (Zappetti et al. 2017). Further details on recent developments and applications of tensegrity structures in robotics can be found in the review by Liu et al. (2022a).

Another field of applications of the tensegrity structures is sensors, actuators and energy absorbers. The ability to vibrate and rapidly transfer loads allowed Sultan and Skelton (1998b) and then Tibert (2002) to conclude that tensegrity structures can be used for absorbing shocks and seismic vibrations, which is desirable in the earthquake-resistant design. The elastic envelope of a tensegrity structure can be used to store kinetic energy in the form of potential elastic energy (Scruggs and Skelton 2006) or as a shape memory actuator (Defossez 2003). Micheletti and colleagues (dos Santos et al. 2015; Micheletti et al. 2018) suggested using shape memory alloy cables for variable-geometry tensegrity structures.

One of the most recent applications of tensegrity structures is materials science. Tensegrity structures were used as elementary cells in generating lattice assemblies with unprecedented mechanical properties (Rimoli and Pal 2017; Rimoli 2018; Salahshoor et al. 2018; Liu et al. 2019). Tensegrity-like lattices were produced at the centimetre scale using additive manufacturing (Liu et al. 2017; Lee et al. 2020; Intrigila et al. 2022) and at the micrometre scale using multiphoton lithography (Vangelatos et al. 2020; Bauer et al. 2021).

In biology, the first research interest in using tensegrity structures for modelling the organization and functioning of living cells was sparked by the striking similarity of tensegrity structures with the architecture of regular viruses (Caspar and Klug 1962) and with the cytoskeleton of a living cell (Ingber 1998). Since 1990s, the tensegrity structures have been extensively used for understanding many biological systems extending from the individual cells (e.g. Ingber 1993, 1998, 2003; Volokh et al.2000; Wang et al. 2001; Wendling et al. 2003; Lazopoulos 2004; Sultan et al. 2004; Lazopoulos and Lazapolou 2005; Gan 2020) to the physiology of mammals (e.g. Levin 2002; Valero-Cuevas et al. 2007; Skelton and de Oliveira 2009).

Part I

Two-dimensional plane tensegrity structures

Chapter 1

Kite structures

The simplest tensegrity structure is a kite that children are playing with for centuries (Figure 1.1).

The simplest kite is composed of two bars and four cables which form a square, as shown in Figure 1.2 in which the thick lines are bars and the thin lines are cables.

Figure 1.1 Child playing with a kite.

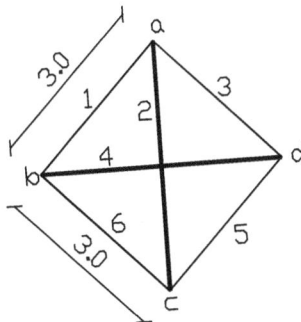

Figure 1.2 Simple square kite.

DOI: 10.1201/9781003370093-2

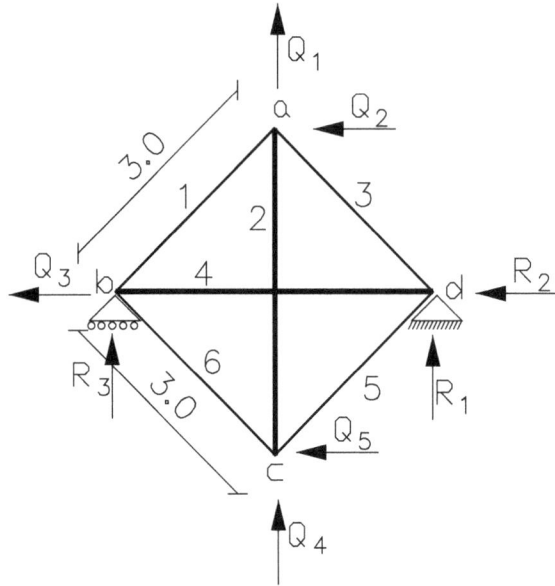

Figure 1.3 Simple tensegrity kite structure.

Each bar is connected at both ends to the cables. The bars overlap each other but they are not connected.

This kite can be used to form the simple tensegrity kite structure shown in Figure 1.3.

In the case where forces F1, F2, F3, F4, F5 and F6 applied to members 1, 2, 3, 4, 5 and 6 equilibrium at nodes a, b and c take the form of a set of equilibrium equations given by Equation 1.1:

$$1.1 \quad \mathbf{AF} = \mathbf{Q}$$

Here

$$1.2 \quad \mathbf{F} = \begin{vmatrix} F1 \\ F2 \\ F3 \\ F4 \\ F5 \\ F6 \end{vmatrix} \quad \mathbf{Q} = \begin{vmatrix} Q1 \\ Q2 \\ Q3 \\ Q4 \\ Q5 \end{vmatrix}$$

Equation 1.1 is defined as the equilibrium equation and matrix **A** the equilibrium matrix.

By considering overall equilibrium of the simple tensegrity kite structure, the magnitude of the reactions R_1, R_2 and R_3 can be determined.

Matrix **A**, the so-called equilibrium matrix, where all forces are assumed to be positive, takes the following form:

$$1.3 \quad \mathbf{A} = 0.5 \begin{vmatrix} \sqrt{2} & 2 & \sqrt{2} & 0 & 0 & 0 \\ -\sqrt{2} & 0 & \sqrt{2} & 0 & 0 & 0 \\ \sqrt{2} & 0 & 0 & 2 & \sqrt{2} & 0 \\ 0 & -2 & 0 & 0 & -\sqrt{2} & -\sqrt{2} \\ 0 & 0 & 0 & 0 & \sqrt{2} & -\sqrt{2} \end{vmatrix}$$

This is a consistent set of equations. There are less equilibrium equations in Equation 1.1 to satisfy than unknown forces. Mechanically, this is an indeterminate simple tensegrity kite structure that can sustain all loadings in this configuration.

Equation 1.1 takes the homogeneous form:

$$1.4 \quad \mathbf{AP} = 0$$

There are more unknown forces P than equations in Equation 1.4. It is possible to prestress the simple tensegrity kite structure. Cable 1 of this indeterminate simple tensegrity kite structure can be prestressed to the magnitude of P_0 and by using Equation 1.4, it is possible to determine the forces induced into the other members of the simple tensegrity kite structure. In this case, **P** takes the following form:

$$1.5 \quad \mathbf{P} = \begin{vmatrix} P1 \\ P2 \\ P3 \\ P4 \\ P5 \\ P6 \end{vmatrix} = P_0 \begin{vmatrix} 1 \\ -\sqrt{2} \\ 1 \\ -\sqrt{2} \\ 1 \\ 1 \end{vmatrix}$$

The fact that prestressing induces compression to the bars and tension to the cables implies that the kite is appropriately prestressed. It can be constructed and prestressed in the given configuration. The prestressing is to the level required to maintain tension in all cables considering expected loads.

If the tensegrity structure is not prestressable and prestressing does not induce tension to the cables and compression to the bars, the proposed tensegrity structure is not feasible. It is actually impossible to construct it in

the proposed configuration. In constructing an unfeasible tensegrity structure, a complete failure or a tensegrity structure with configuration different
from the proposed one can be obtained.

The length of members 1, 2, 3, 4, 5 and 6 of the simple tensegrity kite
structure L1, L2, L3, L4, L5 and L6 takes the following form:

$$1.6 \quad \mathbf{L} = \begin{vmatrix} L1 \\ L2 \\ L3 \\ L4 \\ L5 \\ L6 \end{vmatrix} = 3.0 \begin{vmatrix} 1 \\ \sqrt{2} \\ 1 \\ \sqrt{2} \\ 1 \\ 1 \end{vmatrix}$$

The elements of $\mathbf{P}$ given by Equation 1.5 are in absolute values proportional
to the elements of $\mathbf{L}$ given by Equation 1.6. It is possible to determine the
relative magnitude of the prestressing force by comparing the length of the
relevant members.

It is possible to study graphically the forces induced by prestressing by
using the force diagram shown in Figure 1.4.

The force diagram consists of the equilibrium triangles of the forces acting
at each node. By assuming the prestressing force P_0 in member 1, it is possible
to construct the force diagram of the simple tensegrity kite structure shown
in Figure 1.3, as shown in Figure 1.4. Triangle ABC of the force diagram indicates the equilibrium at node a, triangle ABD at node b, triangle BCD at node
c and triangle ACD at node d. The length of P_i in the force diagram indicates
the magnitude of the force in member i. The arrows indicate the nature of
the forces at the member: tension or compression as shown in Figure 1.5.

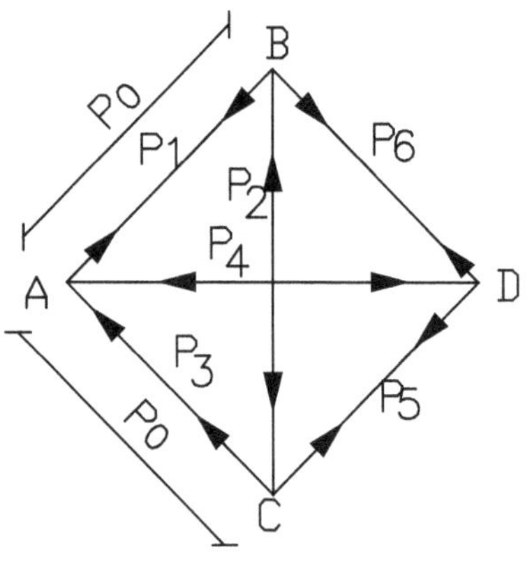

Figure 1.4 Force diagram of the simple tensegrity kite structure shown in Figure 1.2.

TENTION   COMPRESSION

Figure 1.5 Types of forces.

It can be seen that the force diagram is in the shape of the kite and the prestressing forces in absolute values are proportional to the relative length of the members as given by Equation 1.6. It is important to realize that the length of cable 3 is actually proportional to the force induced to cable 6 by prestressing and contrariwise. In this case, it is of no importance since the force induced to cable 3 is identical to the one induced to cable 6.

The fact that it is possible to predict force in all members by assuming the force in one member only indicates that there is one degree of freedom in the prestressing of the simple tensegrity kite structure. The force diagram shows the relative magnitude of the forces induced by prestressing. The precise magnitude of the forces depends on the actual pressing force P_0 applied to one of the elements of the structure, cable or bar which defines the scale of the force diagram.

Also, in the case of a simple tensegrity kite structure based on the shape of a parallelogram kite, the force diagram takes the shape of the kite as shown in Figure 1.6.

It can be seen that the force due to prestressing takes the following form:

$$1.7 \quad \mathbf{P} = \begin{vmatrix} P1 \\ P2 \\ P3 \\ P4 \\ P5 \\ P6 \end{vmatrix} = P_0 \begin{vmatrix} \sqrt{13} \\ -\sqrt{10} \\ 3 \\ -\sqrt{34} \\ \sqrt{13} \\ 3 \end{vmatrix}$$

The length of the members takes the following form:

$$1.8 \quad \mathbf{L} = \begin{vmatrix} L1 \\ L2 \\ L3 \\ L4 \\ L5 \\ L6 \end{vmatrix} = \begin{vmatrix} \sqrt{13} \\ \sqrt{10} \\ 3 \\ \sqrt{34} \\ \sqrt{13} \\ 3 \end{vmatrix}$$

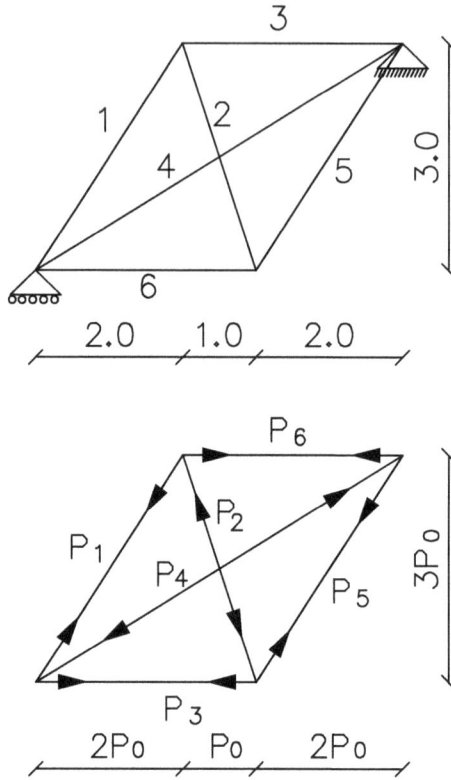

Figure 1.6 Simple tensegrity kite structure in the shape of a parallelogram kite and its force diagram.

Also, in this case, the proportion between the member length and the force induced in it by prestressing is observed. It is important to realize that the length of cable 3 is actually proportional to the force induced to cable 6 by prestressing. This fact is of no importance since the force induced to cable 3 is identical to the one induced to cable 6.

The force diagram in Figure 1.6 implies that the cable net shown in Figure 1.7 is also in equilibrium.

In the case where two cables intersect, they can be connected by four cables 1, 3, 4 and 5, as shown in Figure 1.7. To maintain equilibrium, the net must be designed so that

$$1.9 \quad L_{ad} / L_{bc} = P_2 / P_4$$

Here, L_{ij} is the length between nodes i and j.

A simple tensegrity kite structure in the shape of a trapezium kite and its force diagram are shown in Figure 1.8.

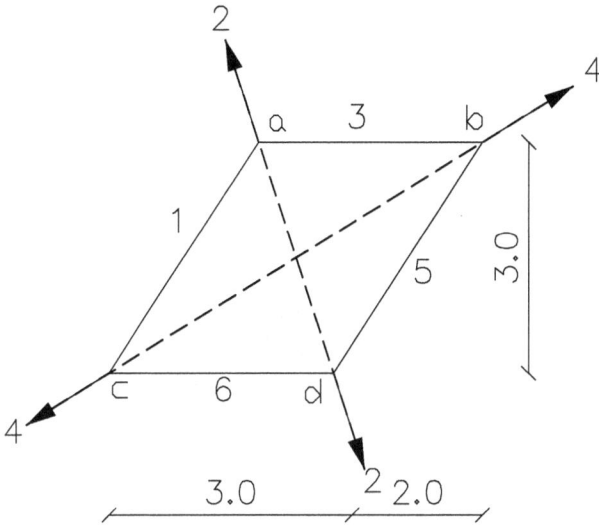

Figure 1.7 Cable net.

Also, in this case, the force diagram takes the shape of the kite. The forces due to prestressing take the following form:

$$1.10 \quad \mathbf{P} = \begin{vmatrix} P1 \\ P2 \\ P3 \\ P4 \\ P5 \\ P6 \end{vmatrix} = 0.5\,P_0 \begin{vmatrix} 14 \\ -\sqrt{153} \\ \sqrt{13} \\ -\sqrt{90} \\ 7 \\ \sqrt{34} \end{vmatrix}$$

The length of the members takes the following form:

$$1.11 \quad \mathbf{L} = \begin{vmatrix} L1 \\ L2 \\ L3 \\ L4 \\ L5 \\ L6 \end{vmatrix} = 0.5 \begin{vmatrix} 7 \\ \sqrt{153} \\ \sqrt{13} \\ \sqrt{90} \\ 14 \\ \sqrt{34} \end{vmatrix}$$

It can be seen that vector $\mathbf{P}$ is proportional to vector $\mathbf{L}$ in absolute values where L3 and L6 interchanged.

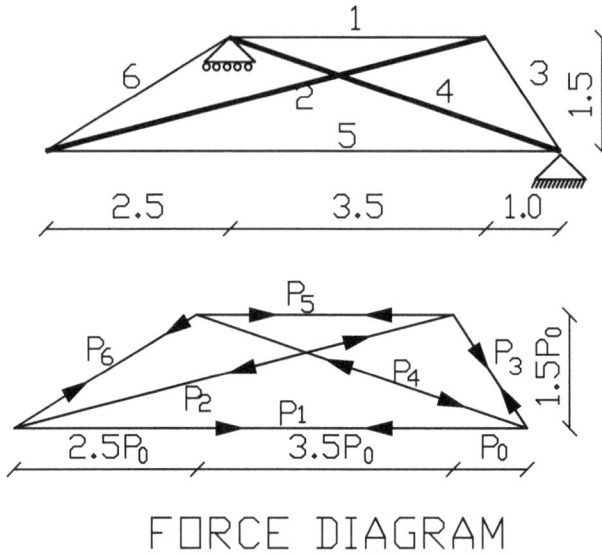

FORCE DIAGRAM

Figure 1.8 Simple trapezium tensegrity kite structure and its force diagram.

It is possible to design simple tensegrity kite structure in very odd shape. For example, the kite shown in Figure 1.9.

In this case, the force diagram shape is completely different from the kite structure shape.

The forces due to prestressing the kite take the following form:

$$1.12 \quad \mathbf{P} = \begin{vmatrix} P1 \\ P2 \\ P3 \\ P4 \\ P5 \\ P6 \end{vmatrix} = 3P_0 \begin{vmatrix} 2\sqrt{5} \\ -\sqrt{17} \\ 1 \\ -\sqrt{13} \\ \sqrt{5} \\ 2\sqrt{2} \end{vmatrix}$$

The length of the members takes the following form:

$$1.13 \quad \mathbf{L} = \begin{vmatrix} L1 \\ L2 \\ L3 \\ L4 \\ L5 \\ L6 \end{vmatrix} = \begin{vmatrix} 3\sqrt{5} \\ 2\sqrt{17} \\ 6 \\ 3\sqrt{13} \\ \sqrt{5} \\ 8\sqrt{2} \end{vmatrix}$$

FORCE DIAGRAM

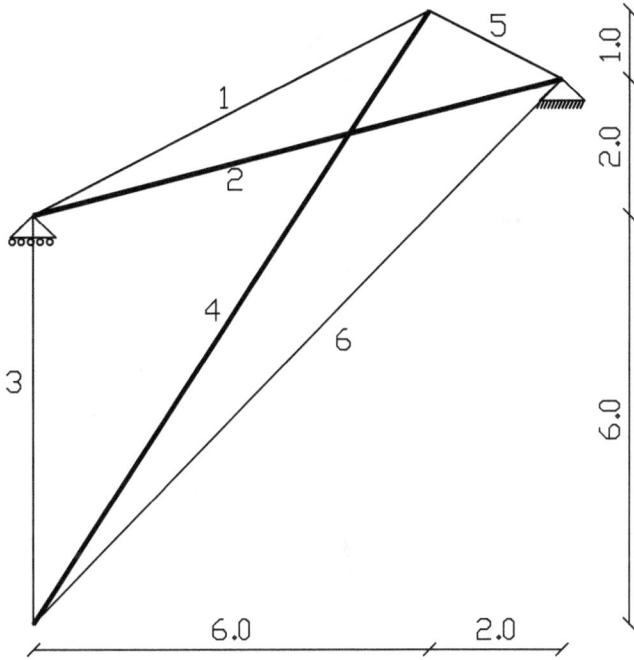

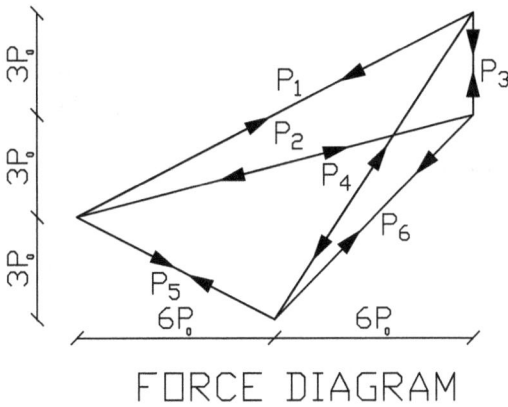

Figure 1.9 Odd shape simple kite structure and its force diagram.

In this case, the relationship between the length of the kite members and the force induced in it by prestressing is not obvious.

A kite may have a more elaborate configuration, as shown in Figure 1.10.

The elaborate kite can be used to construct a tensegrity kite structure shown in Figure 1.11.

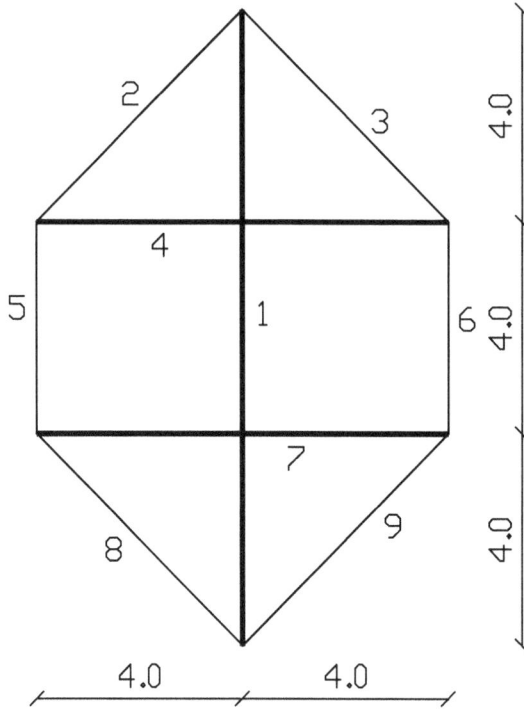

Figure 1.10 Elaborate kite.

The elaborate tensegrity kite structure is composed of three bars and six cables. Each bar is connected to the cables net only. The bars overlap each other but are not connected.

Matrix **A** of this elaborate tensegrity kite structure takes the following form:

$$
1.14 \quad \mathbf{A} = 0.5
\begin{vmatrix}
2 & \sqrt{2} & \sqrt{2} & 0 & 0 & 0 & 0 & 0 & 0 \\
0 & -\sqrt{2} & \sqrt{2} & 0 & 0 & 0 & 0 & 0 & 0 \\
0 & -\sqrt{2} & 0 & 0 & 2 & 0 & 0 & 0 & 0 \\
0 & \sqrt{2} & 0 & 2 & 0 & 0 & 0 & 0 & 0 \\
0 & 0 & -\sqrt{2} & 0 & 0 & 2 & 0 & 0 & 0 \\
0 & 0 & -\sqrt{2} & -2 & 0 & 0 & 0 & 0 & 0 \\
0 & 0 & 0 & 0 & 0 & 0 & 2 & \sqrt{2} & 0 \\
0 & 0 & 0 & 0 & 0 & 0 & 0 & -\sqrt{2} & \sqrt{2} \\
-2 & 0 & 0 & 0 & 0 & 0 & 0 & -\sqrt{2} & -\sqrt{2}
\end{vmatrix}
$$

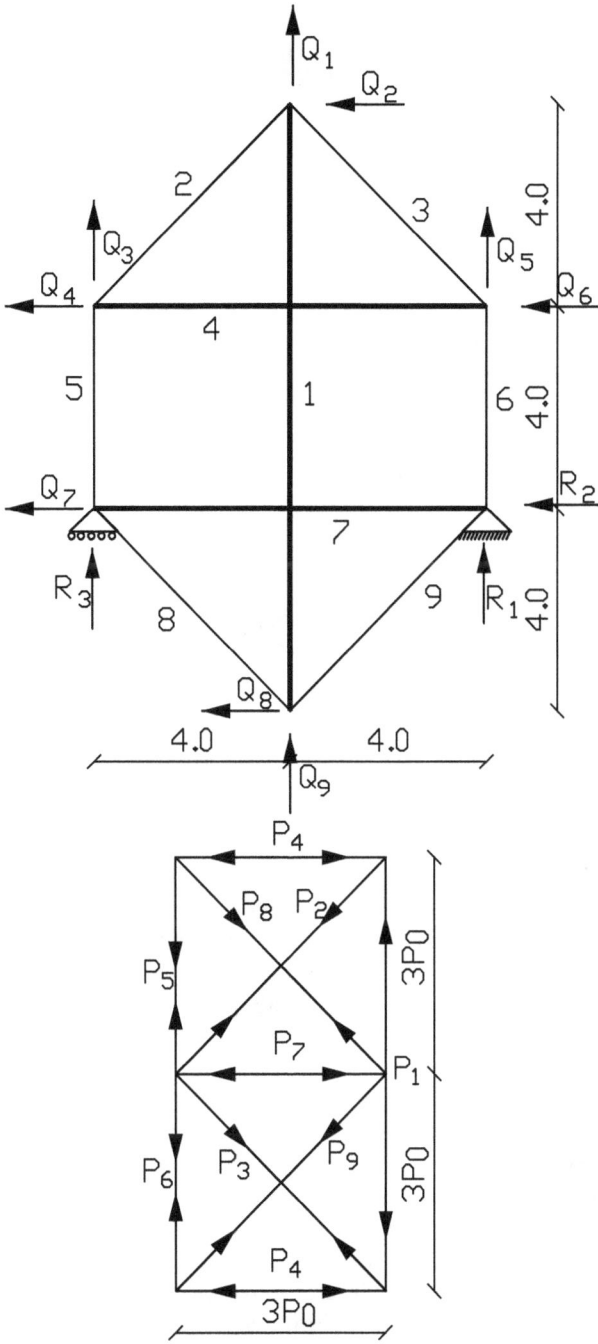

Figure 1.11 Elaborate tensegrity kite structure.

In most cases, the fact that the number of unknowns is equal to the number of equations indicates that the structure is determinate and the forces induced in the members by the load can be determined by considering equilibrium only. But in this case, matrix **A** given by Equation 1.13 is not consistent. Line 6 is a linear combination of lines 2 and 4:

$$1.15 \quad L_6 = -L_2 - L_4$$

Here, L_i is line i of matrix **A**.

The rank of matrix **A** is only 8. Because of the nature of matrix **A**, the homogeneous Equation 1.4 has a solution different from the trivial one: **P** = 0. The elements of **P** of the not trivial solution are the forces induced by prestressing to the members of the tensegrity structure. In the case of the elaborate tensegrity kite structure shown in Figure 1.12 and by using Equation 1.4, **P** takes the following form:

$$1.16 \quad \mathbf{P} = 3P_0 \begin{vmatrix} -2 \\ \sqrt{2} \\ \sqrt{2} \\ -1 \\ 1 \\ 1 \\ -1 \\ \sqrt{2} \\ \sqrt{2} \end{vmatrix}$$

Also, in this case, the prestressing forces can be found by using the force diagram shown in Figure 1.12. Because prestressing induces tension to the cables and compression to the bars, the elaborate tensegrity structure is appropriately prestressed and so it can be constructed in the given configuration.

When this elaborate tensegrity kite structure is loaded, there are not enough equilibrium equations to negotiate with. There will be large nodal displacement partly due to rigid body movement of the members which change matrix **A** until equilibrium is achieved at all the nodes. These large nodal displacements can be observed by the naked eye. The method of analysis of these displacements is presented in Vilnay (1990). Structures with these properties are defined mechanically infinitesimal mechanism. Most of the tensegrity structures considered in literature belong to this category.

A very interesting feature of the infinitesimal mechanism is the so-called fitted load. The tensegrity structure can sustain "fitted load" in its prestressed configuration, the nodal displacements in this loading case are small, similar

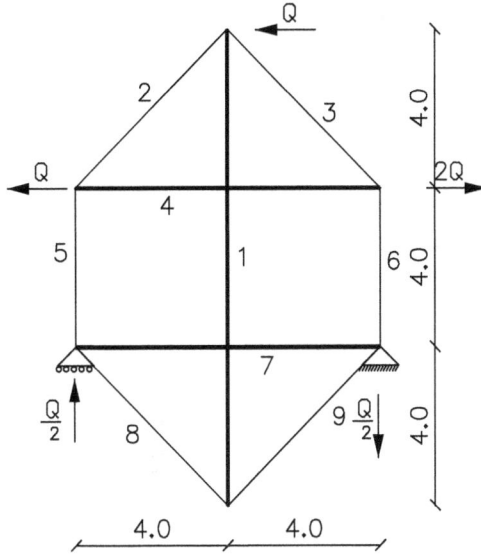

Figure 1.12 "Fitted load".

in magnitude to the displacements of loaded determinate or indeterminate structures.

In the case of the elaborate tensegrity kite shown in Figure 1.11, the "fitted load" $\mathbf{Q}_f$ can be found by using Equation 1.15, which takes the following form:

$$1.17 \quad \mathbf{Q}_f = \begin{vmatrix} Q1 \\ Q2 \\ Q3 \\ Q4 \\ Q5 \\ -Q2-Q4 \\ Q7 \\ Q8 \\ Q9 \end{vmatrix}$$

In the case where the load is "fitted load", following Equation 1.17, Equation 1.1 is a consistent set of linear equations which indicates that the tensegrity structure can sustain "fitted load" in its prestressed configuration and no change of matrix **A** is required. Methods of analysing the nodal displacements and the forces induced by "fitted load" are presented in Vilnay (1990).

This method of establishing the conditions of "fitted load" requires an extensive cumbersome and tedious study of matrix **A**.

In most practical cases, it is necessary to determine if a given load is "fitted load". In these cases, the so-called mechanical method can be used. If it is possible to identify members of the tensegrity structure and establish equilibrium considering forces in them and the given load, the load is "fitted load". For example, the load shown in Figure 1.12 is "fitted load".

It can be seen that the forces **F** given by Equation 1.18 induced to members 2, 3, 4, 5 and 6 shown in full lines in Figure 1.12 are in equilibrium with the given load.

$$1.18 \quad \mathbf{F} = Q/2 \begin{vmatrix} 0 \\ -\sqrt{2} \\ \sqrt{2} \\ 3 \\ -1 \\ 1 \\ 0 \\ 0 \\ 0 \end{vmatrix}$$

So the load shown in Figure 1.12 is "fitted load" which is consistent with the condition given by Equation 1.17.

The disadvantage of the "mechanical method" is that it requires a thorough understanding of the tensegrity structure.

A more mechanical method that can be used to define "fitted load" is the so-called "equilibrium method". When the "mechanical method" is used, prestressing of the tensegrity structure and the loading are carried out simultaneously. If it is possible to obtain a consistent force diagram, the load is "fitted load". For example, the load with its relevant reactions applies to the elaborate tensegrity kite structure shown in Figure 1.11 is shown in Figure 1.13.

It is assumed that the elaborate tensegrity kite structure is prestressed by inducing $6P_0$ to bar 1, as shown in Figure 1.11. If it is possible to get a consistent force diagram considering the force of $6P_0$ in bar 1 and the given load, the load is "fitted load". The consistent force diagram in Figure 1.13 implies that the load shown in Figure 1.13 is "fitted load". The advantage of this method is that it is very mechanical and a thorough examining of the tensegrity structure is not required. The force diagram in Figure 1.13 displays the forces in the tensegrity structure only when the prestressing and the loading are carried out simultaneously. This is not the case where the tensegrity

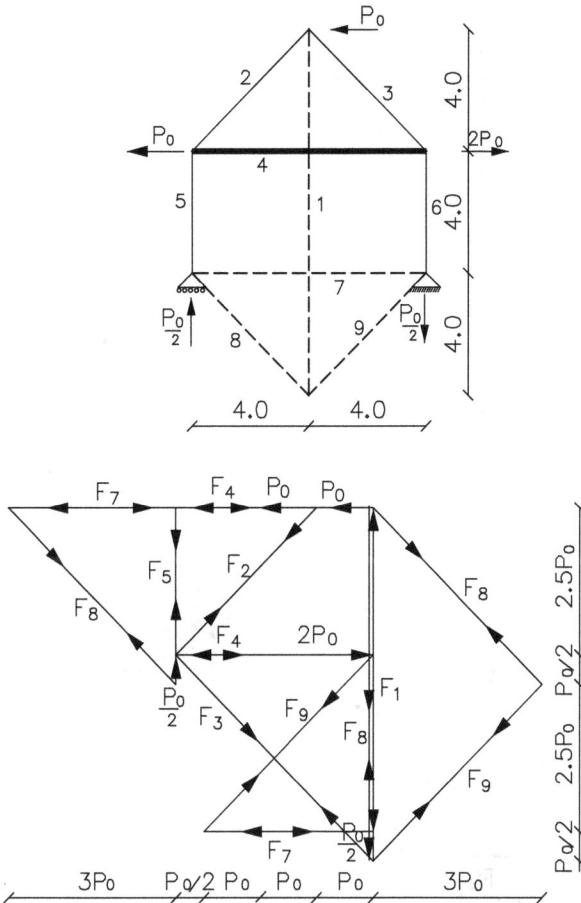

Figure 1.13 Elaborate tensegrity kite structure loaded.

structure is prestressed and then loaded. In this case, the load changes the prestressing forces. The methods of analysis can be found in Vilnay (1990).

Both the "mechanical method" and the "equilibrium method" are practically the same. Both examine if the tensegrity structure can be in equilibrium with the given load. The "mechanical method" examines it straightforward. The "equilibrium method" is based on the fact that as the prestressing forces are in equilibrium, so if the prestressing forces and the proposed load are in equilibrium with the forces in the tensegrity structure, the proposed load by itself should be in equilibrium too.

The elaborate tensegrity kite structure can take the odd configuration shown in Figures 1.14 and 1.15.

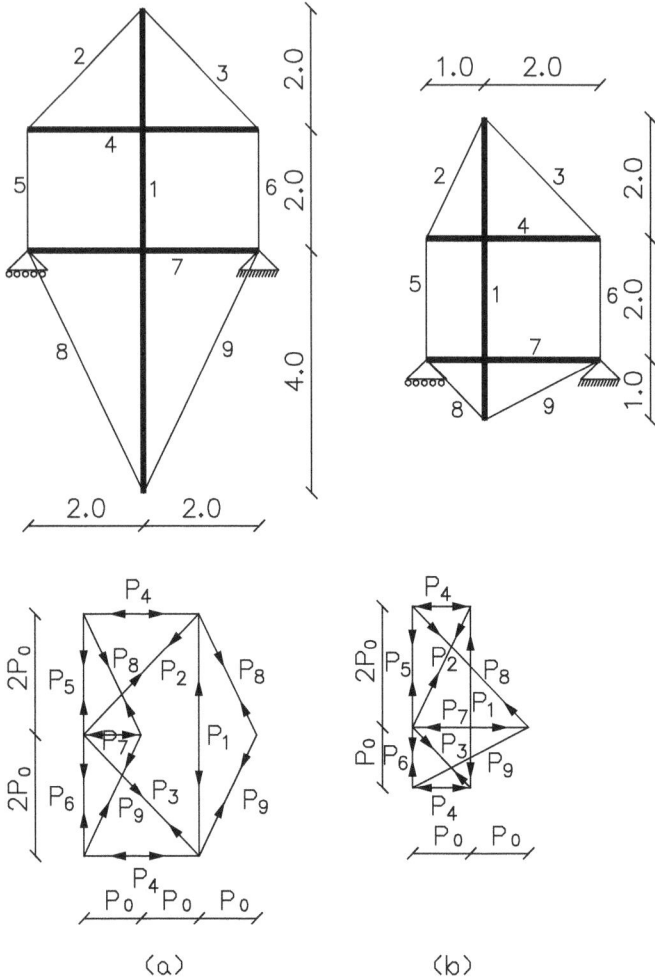

Figure 1.14 Odd elaborate tensegrity kite structure.

The odd elaborate tensegrity kite structures shown in Figures 1.14 and 1.15 are associated with a consistent force diagrams. In all cases, tension was induced to the cables and compression to the bars. These elaborate tensegrity kite structures are appropriately prestressable and so they are feasible tensegrity structures and can be constructed in the given configuration.

Not all elaborate tensegrity kite structures are feasible and can be constructed in the proposed configuration. For example, the elaborate tensegrity kite shown in Figure 1.16.

It is impossible to establish a consistent prestressing force diagram as it is shown in Figure 1.16. Cable 6 is not in the "right" direction. This is an

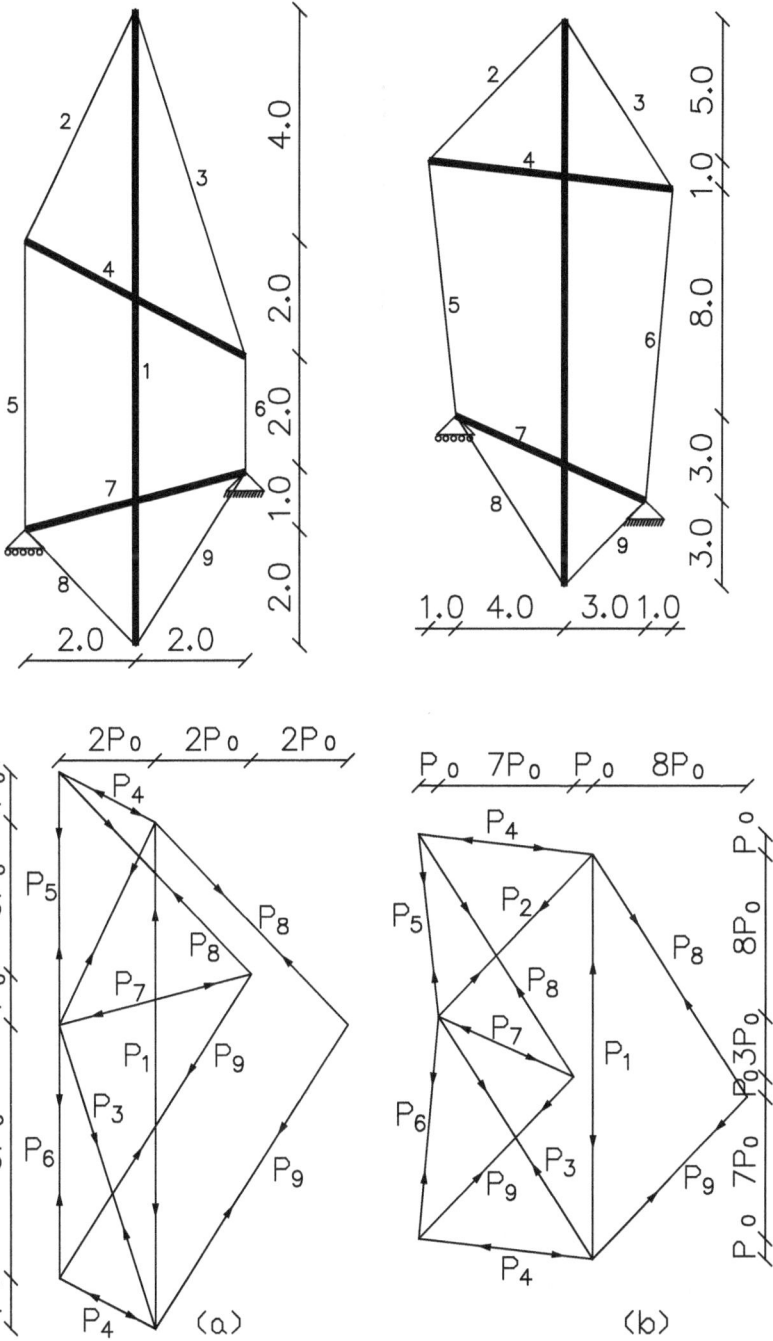

Figure 1.15 Odd elaborate tensegrity kite structure.

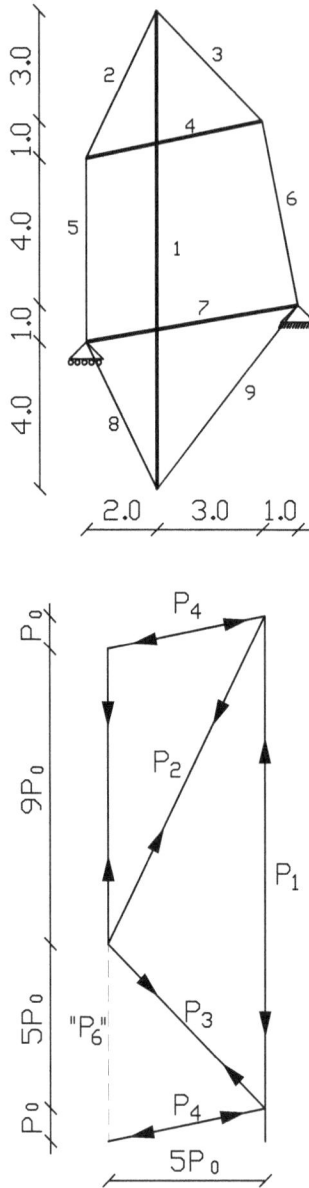

Figure 1.16 Proposed elaborate tensegrity kite structure.

indication that this elaborate tensegrity kite structure is not feasible and cannot be constructed in this configuration. An attempt to build it in this configuration may, if care is taken, result with different configuration or in a complete failure.

Chapter 2

Tensegrity nets

2.1 CONFIGURATION OF TENSEGRITY NETS

The methods used to construct simple kites can be used to form tensegrity nets. Also, tensegrity nets are composed of cables and bars. Each bar is connected at both ends to the cables nodes. In the case of plane tensegrity nets, the bars may overlap each other but they are not connected to each other. A feasible tensegrity net should be appropriately prestressable. It should be prestressable and prestressing should induce tension to the cables and compression to the bars.

A very simple tensegrity net, tensegrity net A, is shown in Figure 2.1

The fact that the net is prestressable can be examined by assuming the force in one member of the tensegrity net and checking if equilibrium can be maintained throughout the net. A simple method of checking is to use a force diagram. For example, in the case of the net shown in Figure 2.1, it is

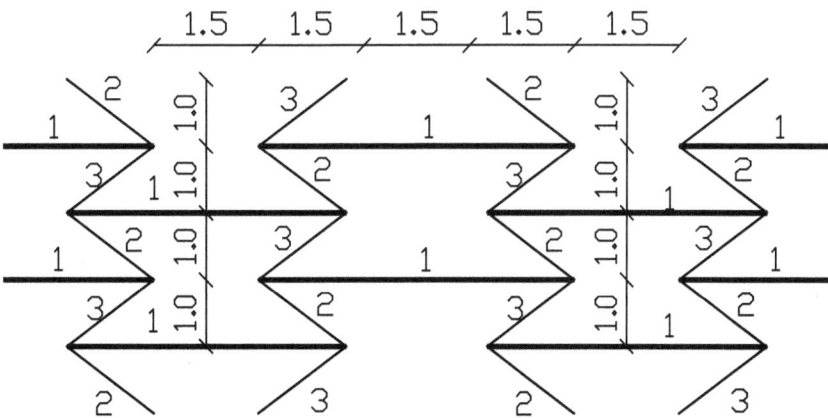

Figure 2.1 Very simple tensegrity net A.

DOI: 10.1201/9781003370093-3

possible to assume the force in bar 1; the forces induced in cables 2 and 3 can be found by using the force diagram shown in Figure 2.2. The scale of the force diagram indicates the actual magnitude of the forces.

A more elaborate tensegrity net, tensegrity net B, is shown in Figure 2.3.

Overall equilibrium implies $F_1 = -F_4$ and the force diagram shown in Figure 2.3 indicates the forces induced to members 2 and 3 due to prestressing.

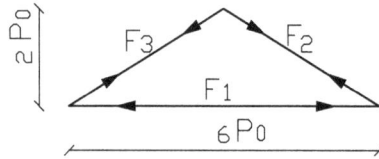

Figure 2.2 Force diagram of the tensegrity net A shown in Figure 2.1.

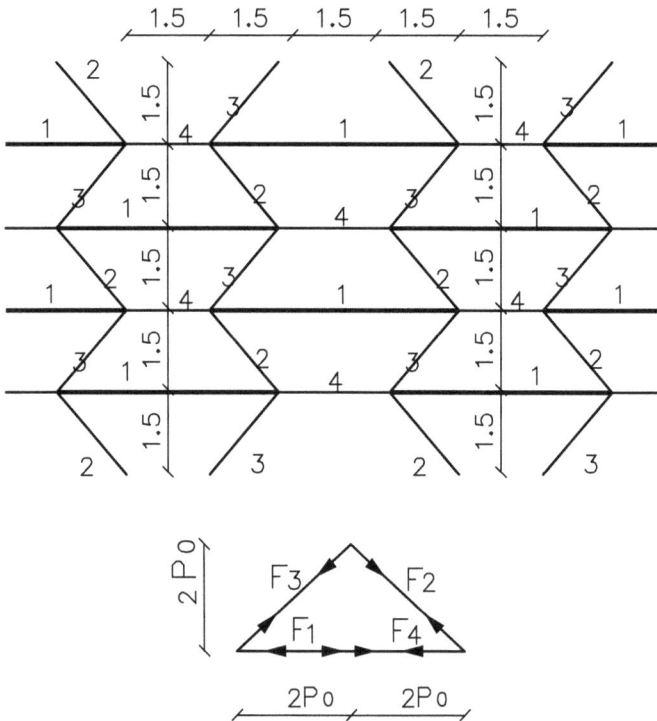

Figure 2.3 Tensegrity net B.

Another type of a tensegrity net, tensegrity net C, is shown in Figure 2.4.

In the general case of prestressing where the prestressing forces vary from one side of the tensegrity cell to the other, tensegrity net C prestressing forces are shown in Figure 2.5.

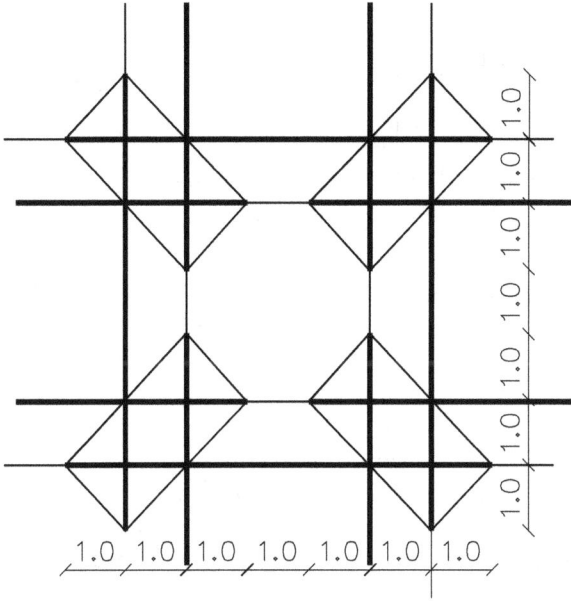

Figure 2.4 Tensegrity net type C.

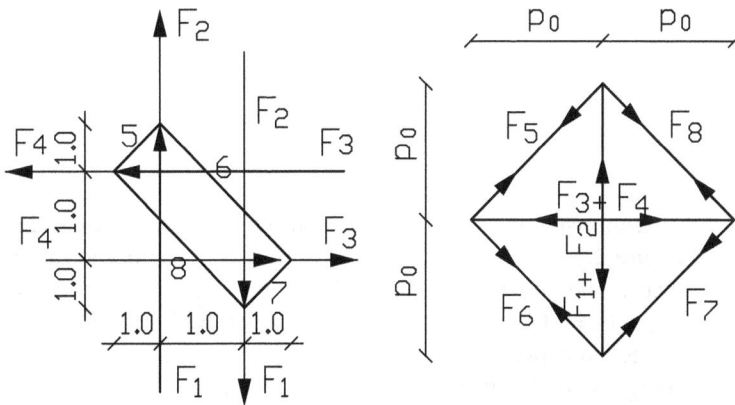

Figure 2.5 Tensegrity net C prestressing forces.

Equilibrium at the nodes implies:

2.1
$$
\begin{aligned}
F_1 + F_2 - F_5\sqrt{2}/2 - F_6\sqrt{2}/2 &= 0 \quad &\text{L1}\\
F_6\sqrt{2}/2 - F_5\sqrt{2}/2 &= 0 \quad &\text{L2}\\
F_3 + F_4 - F_5\sqrt{2}/2 - F_8\sqrt{2}/2 &= 0 \quad &\text{L3}\\
F_5\sqrt{2}/2 - F_8\sqrt{2}/2 &= 0 \quad &\text{L4}\\
F_1 + F_2 - F_7\sqrt{2}/2 - F_8\sqrt{2}/2 &= 0 \quad &\text{L5}\\
F_7\sqrt{2}/2 - F_8\sqrt{2}/2 &= 0 \quad &\text{L6}\\
F_3 + F_4 - F_6\sqrt{2}/2 - F_7\sqrt{2}/2 &= 0 \quad &\text{L7}\\
F_6\sqrt{2}/2 - F_7\sqrt{2}/2 &= 0 \quad &\text{L8}
\end{aligned}
$$

It can be seen that Equation 2.1 yields the following three independent equations:

2.2
$$
\begin{aligned}
F_5 &= F_6 = F_7 = F_8\\
F_1 + F_2 &= \sqrt{2}F_5\\
F_3 + F_4 &= \sqrt{2}F_5
\end{aligned}
$$

It can be seen that the designer has three degrees of freedom to determine the prestressing forces. The magnitude of F_5 and accordingly F_1 or F_2 and F_3 or F_4, taking care to maintain tension in all cables.

In the case where F_5 is assumed to be $P_0\sqrt{2}$, Equation 2.2 takes the following form:

2.3
$$
\begin{aligned}
F_5 &= P_0\sqrt{2}\\
F_6 &= F_7 = F_8 = P_0\sqrt{2}\\
F_1 + F_2 &= 2P_0\\
F_3 + F_4 &= 2P_0
\end{aligned}
$$

A graphical demonstration of these facts can be seen in the force diagram shown in Figure 2.5. It can be seen that the force diagram implies that where P_5 is assumed to be $P_0\sqrt{2}$, the magnitude of F_6, F_7 and F_8 also equates to $P_0\sqrt{2}$ and $F_1 + F_2 = 2P_0$ and $F_3 + F_4 = 2P_0$ as given by Equation 2.3. In the case of homogeneous prestressing, $F_1 = F_2 = F_3 = F_4 = P_0$.

Thus, tensegrity net C can be appropriately prestressable and so it can be constructed in the configuration shown in Figure 2.4.

A very similar tensegrity net D is shown in Figure 2.6.

The general case of prestressing tensegrity net D is shown in Figure 2.7.

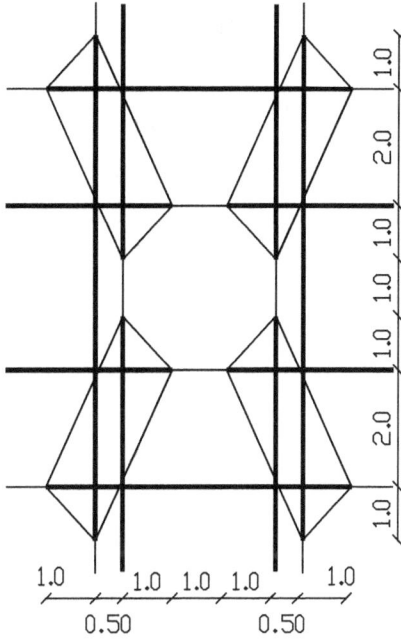

Figure 2.6 Tensegrity net D.

Equilibrium at the nodes takes the following form:

$$2.4 \quad \begin{aligned} F_1 + F_2 - F_5\sqrt{2}/2 - F_6\cos\alpha &= 0 \\ F_5\sqrt{2}/2 - F_6\sin\alpha &= 0 \qquad F_6 = \sqrt{2}F_5/(2\sin\alpha) \\ F_3 + F_4 - F_5\sqrt{2}/2 - F_8\sin\alpha &= 0 \\ F_5\sqrt{2}/2 - F_8\cos\alpha &= 0 \qquad F_8 = \sqrt{2}F_5/(2\cos\alpha) \\ F_1 + F_2 - F_7\sqrt{2}/2 - F_8\cos\alpha &= 0 \\ F_7\sqrt{2}/2 - F_6\sin\alpha &= 0 \qquad F_7 = \sqrt{2}F_5/(2\cos\alpha) \\ F_3 + F_4 - F_6\sqrt{2}/2 - F_7\sqrt{2}/2 &= 0 \\ -F_7\sqrt{2}/2 + F_6\cos\alpha &= 0 \qquad (\cos\alpha)^2 = (\sin\alpha)^2 \end{aligned}$$

It can be seen that the last equation of Equation 2.4 indicates that prestressing is possible only where $\alpha = 45°$. When $\alpha = 45°$, tensegrity net D takes the configuration of tensegrity net C shown in Figure 2.4. This fact is also demonstrated in the force diagram shown in Figure 2.7. The force diagram of the tensegrity net D is not consistent when the net takes the configuration shown in Figure 2.6. In an attempt to prestress tensegrity net D, the net, if appropriate care is taken, can take tensegrity net C configuration or may be a complete failure.

F2 F2

F4   5

F3   1.0

6

8   2.0

F4   F3   1.0

7

F1 F1

1.0   1.0
0.5

P0 P0

F5

P0 F8

2P0 F6   F1+F2

F7

should be horizontal
equale to F3+F4

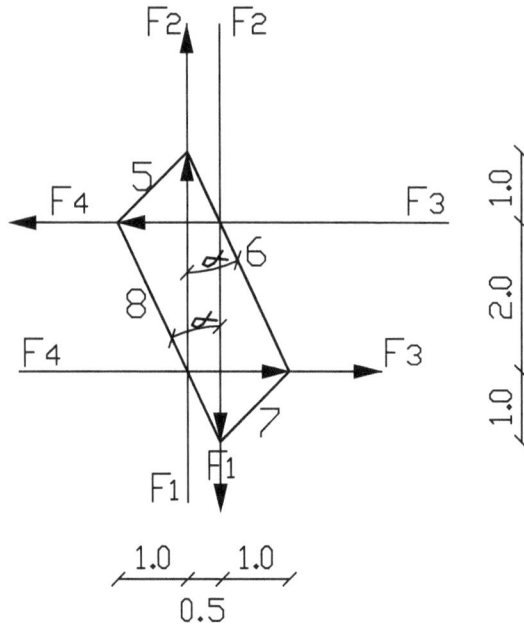

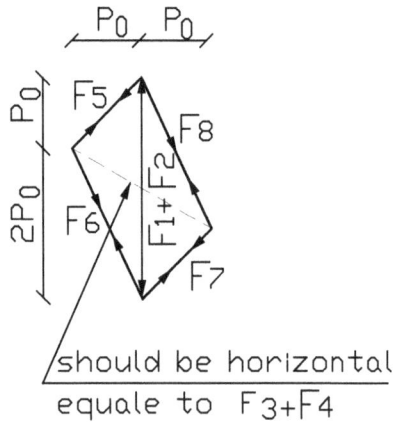

FORCE DIAGRAM

Figure 2.7 Tensegrity net D prestressing forces.

Another version of the tensegrity net C, tensegrity net E, is shown in Figure 2.8.

The general case of prestressing forces of tensegrity net E is shown in Figure 2.9.

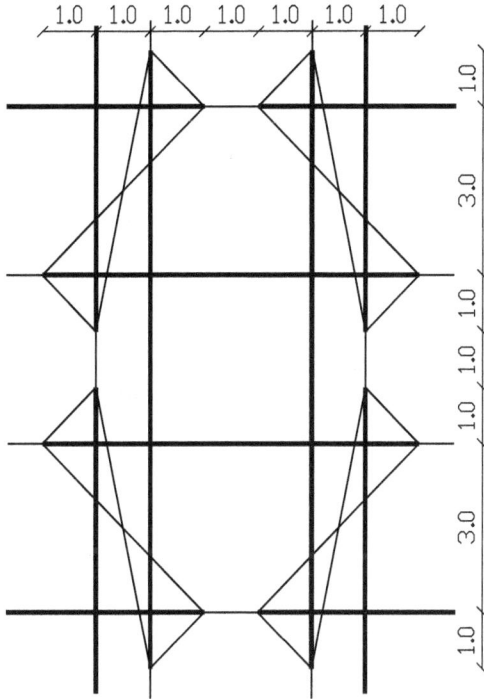

Figure 2.8 Tensegrity net E.

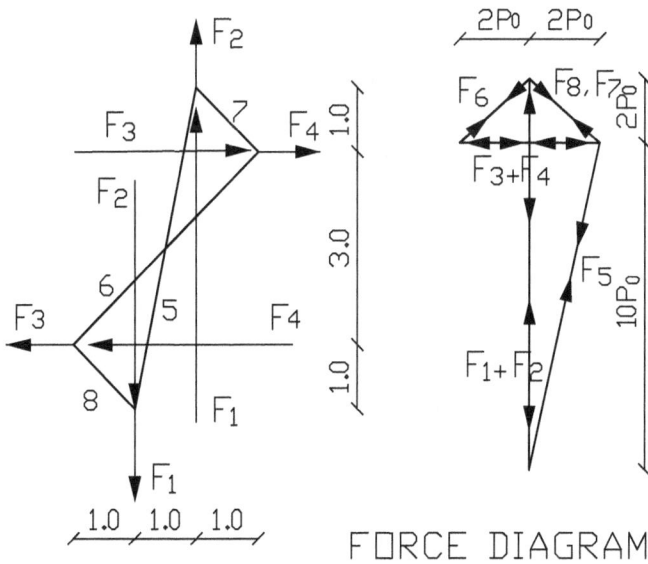

Figure 2.9 Prestressing tensegrity net E.

It can be seen that overall equilibrium of tensegrity net E is satisfied where

$$F_6 = F_7 = F_8 = 2\sqrt{2}P_0$$
$$2.5 \quad F_1 + F_2 = 12P_0$$
$$F_3 + F_4 = 4P_0$$

Also, in this case, the designer has three degrees of freedom to determine the prestressing forces.

In examining the feasibility of a tensegrity structure, if it is possible to show a singular case of appropriately prestressing force scheme, it is enough to conclude that this is a feasible tensegrity structure. Predicting that a proposed tensegrity structure is not feasible should be based on a thorough study of all possible prestressing force schemes.

In the case of symmetrical tensegrity nets, it is easy to assume a possible symmetrical prestressing. For example, in the case of the symmetrical tensegrity net type F shown in Figure 2.10, a possible symmetrical pattern of prestressing forces is shown in Figure 2.11.

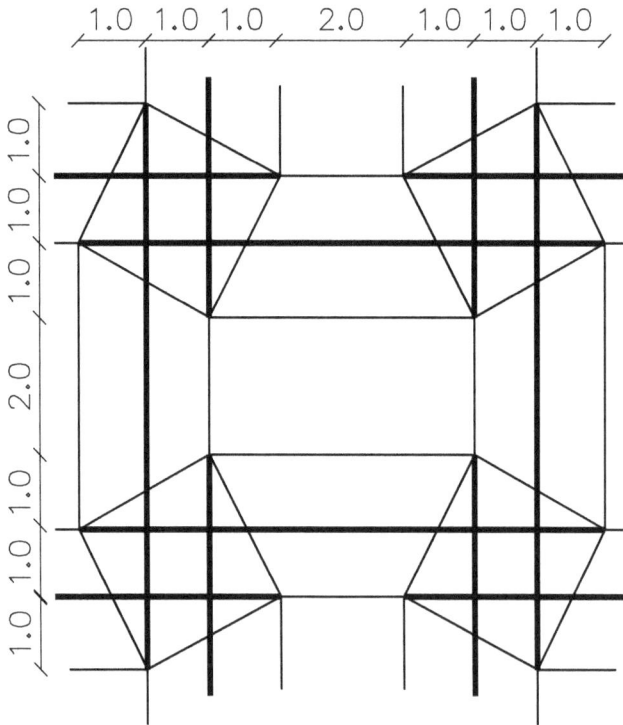

Figure 2.10 Tensegrity net F.

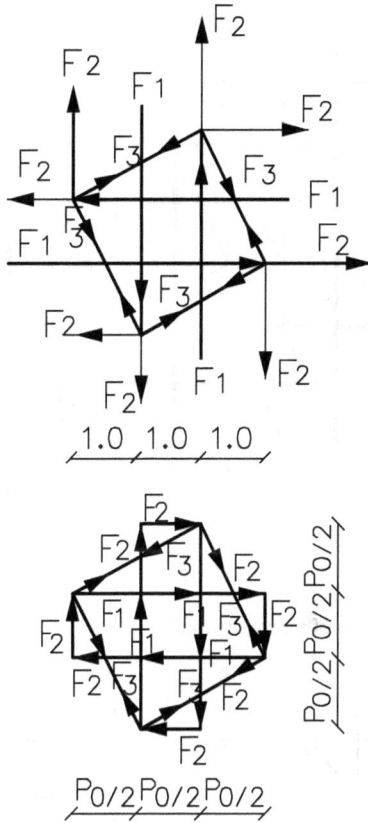

Figure 2.11 Prestressing forces of tensegrity net F.

Overall equilibrium implies:

$$2.6 \quad \begin{array}{l} F_1 = 2F_2 \\ F_1 = P_0, F_2 = P_0 / 2 \end{array}$$

The fact that tensegrity net F can be appropriately prestressed indicates that it is a feasible tensegrity structure and can be constructed in this configuration. Determine a possible prestressing pattern does not indicate that it is the only possible one. The actual prestressing forces depends on the method of prestressing as described in Section 2.2.

Another version of tensegrity net F is shown in Figure 2.12.

Not always is it easy to determine the prestressing forces. In some cases ingenious methods should be used to determine the prestressing forces. An example of these methods is shown in Figure 2.13 where tensegrity net is considered.

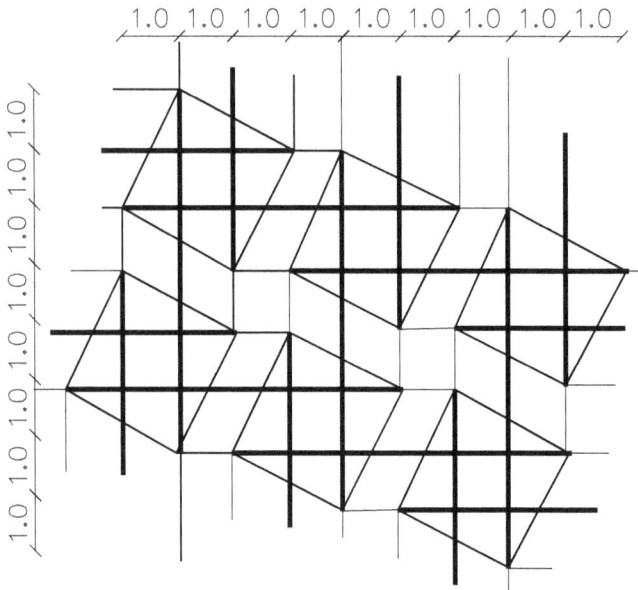

Figure 2.12 Different version of tensegrity net F.

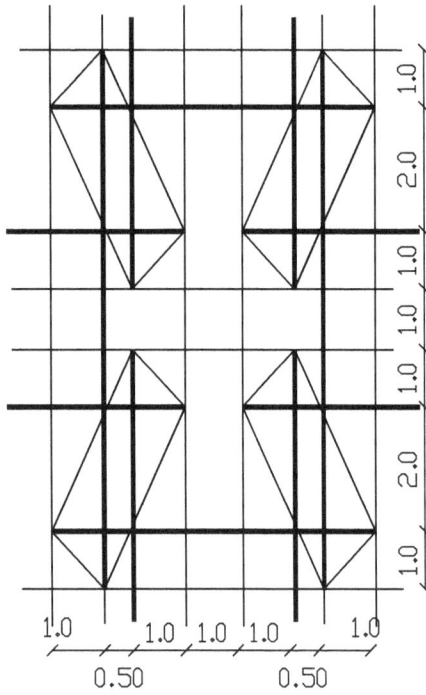

Figure 2.13 Tensegrity net G.

Symmetrical prestressing forces of tensegrity net G are shown in Figure 2.14.

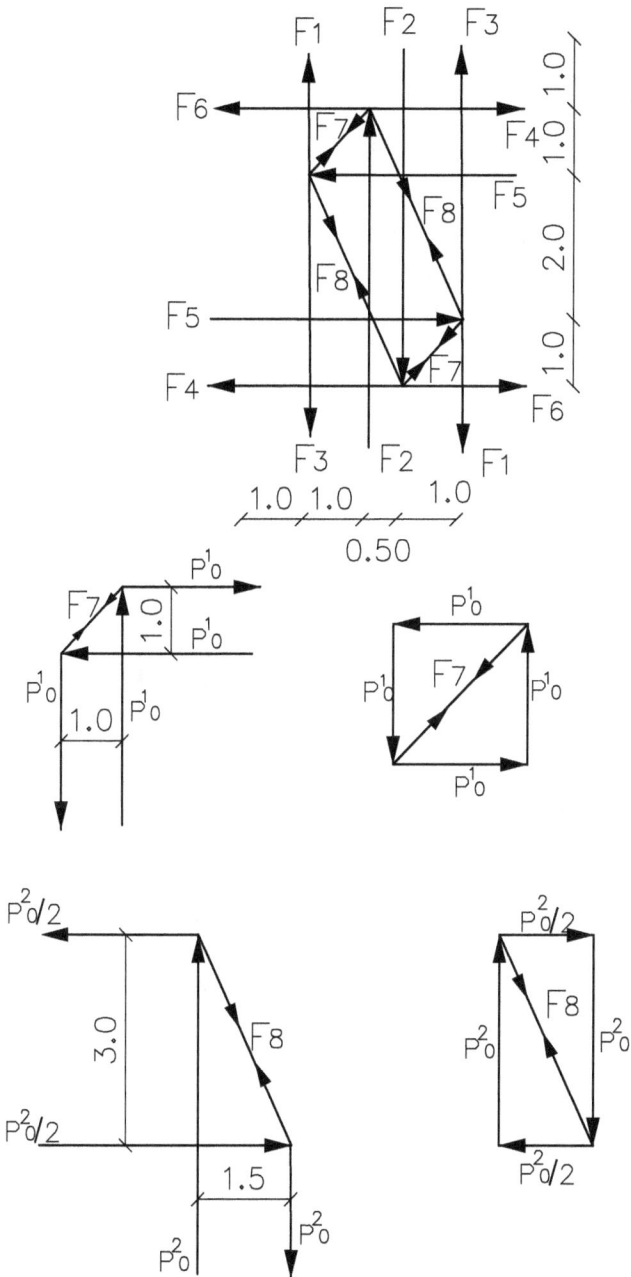

Figure 2.14 Symmetrical prestressing of tensegrity net G.

Formation of equilibrium systems out of the tensegrity net elements can be used to predict the required relationship between the prestressing forces as shown in Figure 2.14. It can be seen that in this symmetrical prestressing, there are two degrees of freedom P_0^1 and P_0^2. It is up to the designer to determine the magnitude of these prestressing forces.

$$2.7 \quad \begin{aligned} &F_1 = P_0^2, F_2 = P_0^1 + P_0^2, F_3 = P_0^1, F_4 = P_0^1, F_5 = P_0^1 + P_0^2/2, \\ &F_6 = P_0^2/2, F_7 = \sqrt{2}P_0^1, F_8 = \sqrt{5}P_0^2/2 \end{aligned}$$

In any case, tensegrity net G is appropriately prestressable and so it is a feasible tensegrity net.

Not all proposed tensegrity nets are feasible; for example, the tensegrity net H shown in Figure 2.15.

The general prestressing forces are shown in Figure 2.16

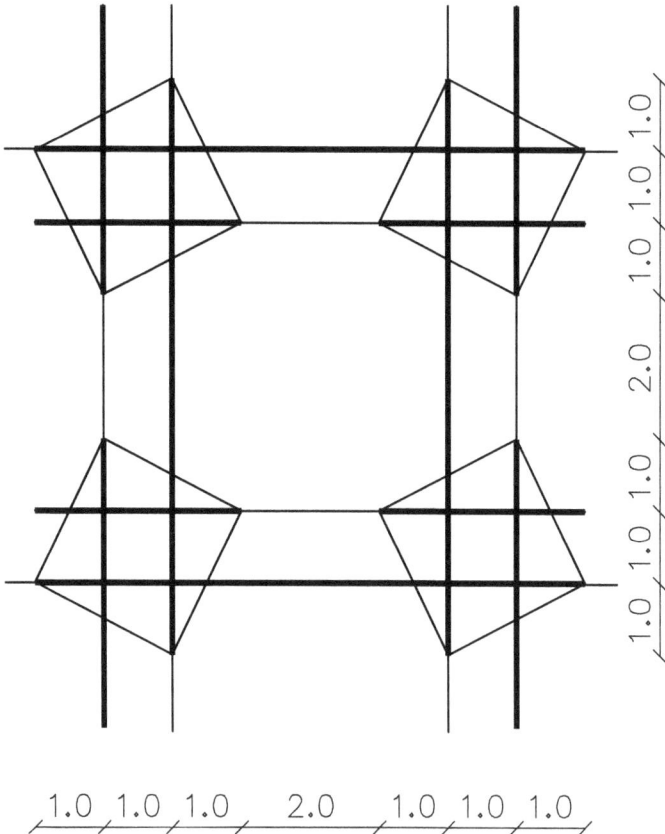

Figure 2.15 Tensegrity net H.

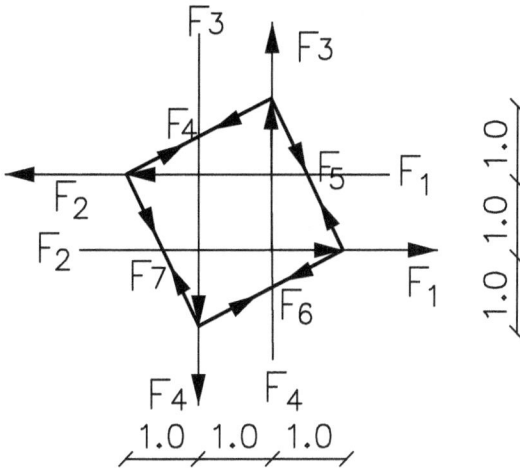

Figure 2.16 Prestressing tensegrity net H.

It can be seen that equilibrium implies an unbalanced moment M when the bars are in compression and the cables are in tension.

$$2.8 \quad M = F_1 + F_2 + F_3 + F_4$$

Because the prestressing forces are not in equilibrium, this net is not a feasible tensegrity net in this configuration. In an attempt to prestress tensegrity net H, the tensegrity, if appropriate care is taken, may take the form of the tensegrity net shown in Figure 2.17, where during prestressing, the symmetry of the structure and the length of the cables are maintained. In the case when other restrains are applied during prestressing, different configurations may be obtained. Anyway, it is very difficult to apply any restraint during prestressing and it is most likely that any attempt to prestress the tensegrity net H will end up with a complete failure.

2.2 PRESTRESSING TENSEGRITY NETS

A tensegrity net gains its rigidity by prestressing. Prestressing should induce tension to the cables and compression to the bars. In large tensegrity nets, methodical methods of prestressing to ensure that proper forces are induced to all members of the tensegrity net must be adopted. In the process of prestressing, it is assumed that the cables are slack and loose and they are straight and tight after prestressing only.

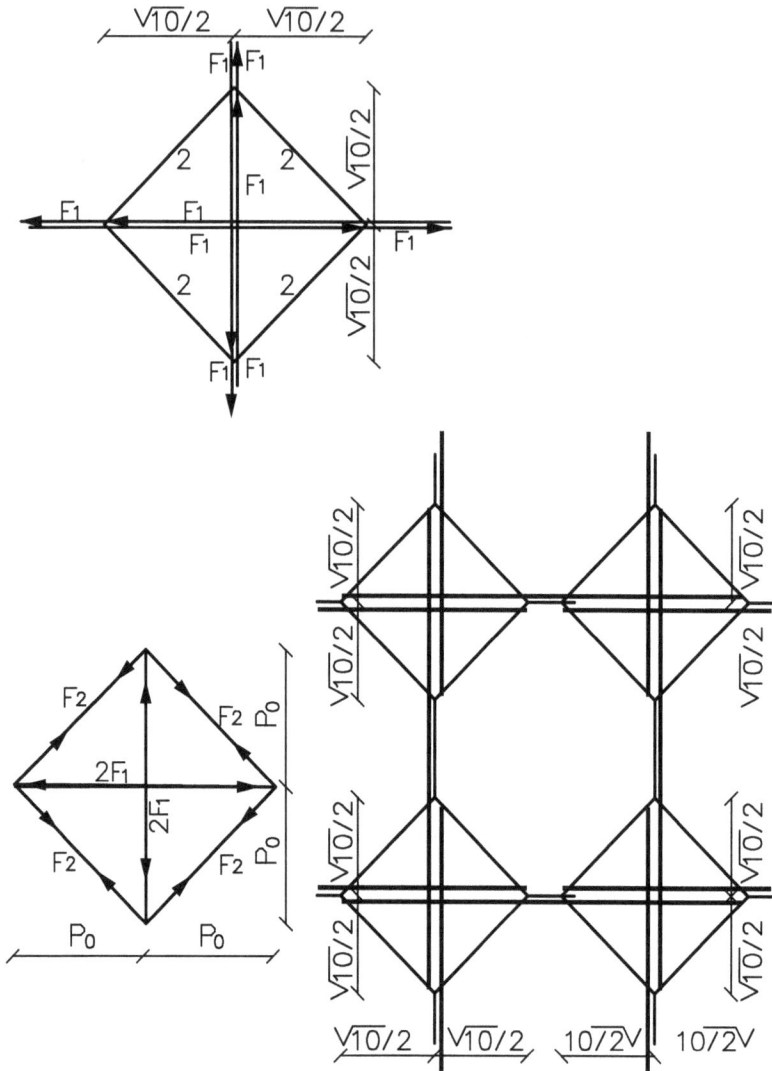

Figure 2.17 Possible configuration of the prestressed tensegrity net H.

Methods of prestressing are up to the designer of the net. Examples of prestressing methods are presented in this chapter.

For example, in the case of the tensegrity net shown in Figure 2.3, it is proposed to prestress the tensegrity net in sections. The first section is shown in Figure 2.18. It is prestressed by inducing force $\sqrt{2}P_0$ to cable a. After the first section is prestressed, a second section shown in Figure 2.18 is added.

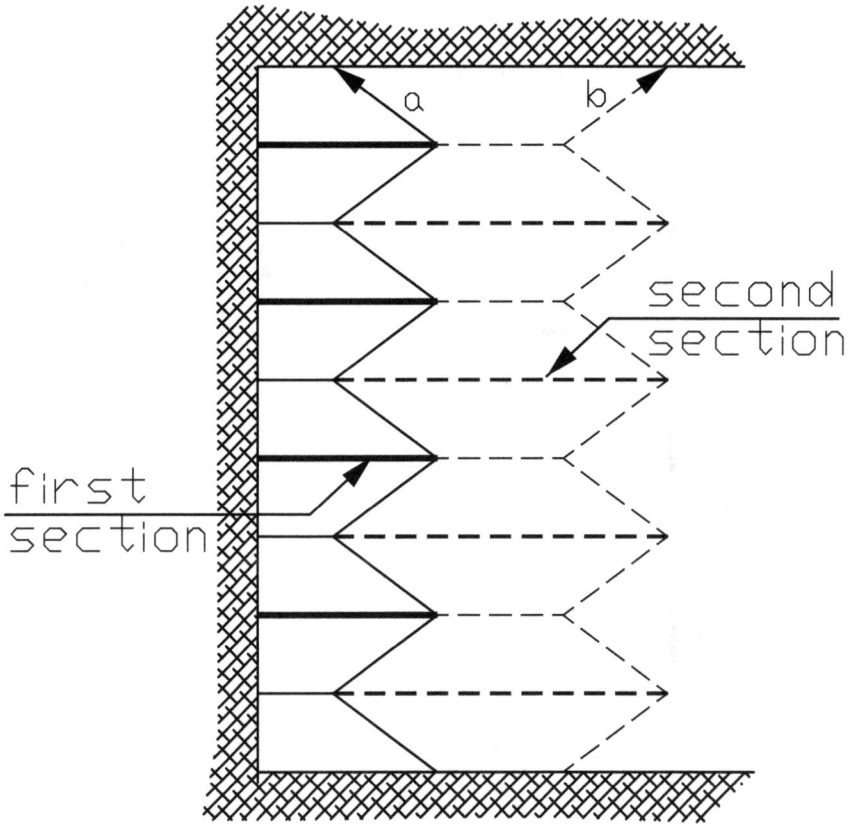

Figure 2.18 Prestressing the tensegrity net shown in Figure 2.3.

At this stage, cables a and b are prestressed to $\sqrt{2}P_0$ simultaneously. At the end of this stage, cable a is prestressed by $2\sqrt{2}P_0$ and cable b by $\sqrt{2}P_0$. Then another section is added and again the same procedure is followed until all necessary sections are added to form the tensegrity net.

In the case of the tensegrity net shown in Figure 2.8, the basic section takes the form shown in Figure 2.19.

The basic section can be prestressed by inducing the designed force F_1 as shown by the force diagram in Figure 2.19. If possible, the best way to prestress the tensegrity net is to prestress all basic sections simultaneously in one go. This requires a certain level of technology and it ensures full control on the forces induced to the tensegrity net members. If this method is not applicable, the prestressing can be carried out section by section. At the first stage of prestressing, the first basic section is prestressed by F_1, then an additional section is added. This section may be prestressed of F_1'. Since

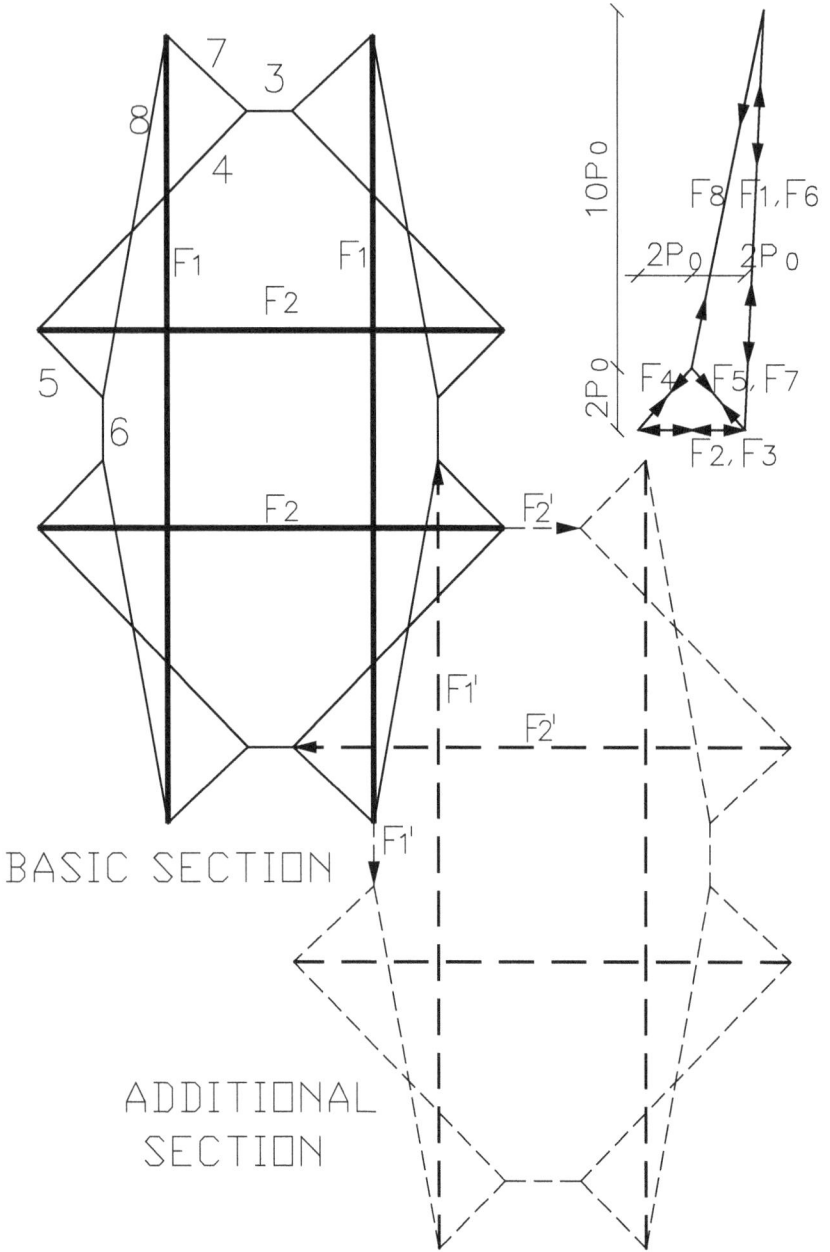

Figure 2.19 The basic section of the tensegrity net shown in Figure 2.8.

the basic section is in equilibrium with F_1', as shown in the force diagram in Figure 2.9, this load is "fitted load". F_1' changes the forces in the elements of the first basic section, but the change of its configuration is marginal. It is in the order of magnitude of the elastic deformation of common structures. The change in the forces in the elements of the basic section can be found by using the methods of analysis presented in Vilnay (1990). By following this method, another additional sections can be added until the whole tensegrity is assembled and prestressed.

The same methods can be used in the case of the tensegrity net shown in Figure 2.10. In this case, the basic section takes the form shown in Figure 2.20.

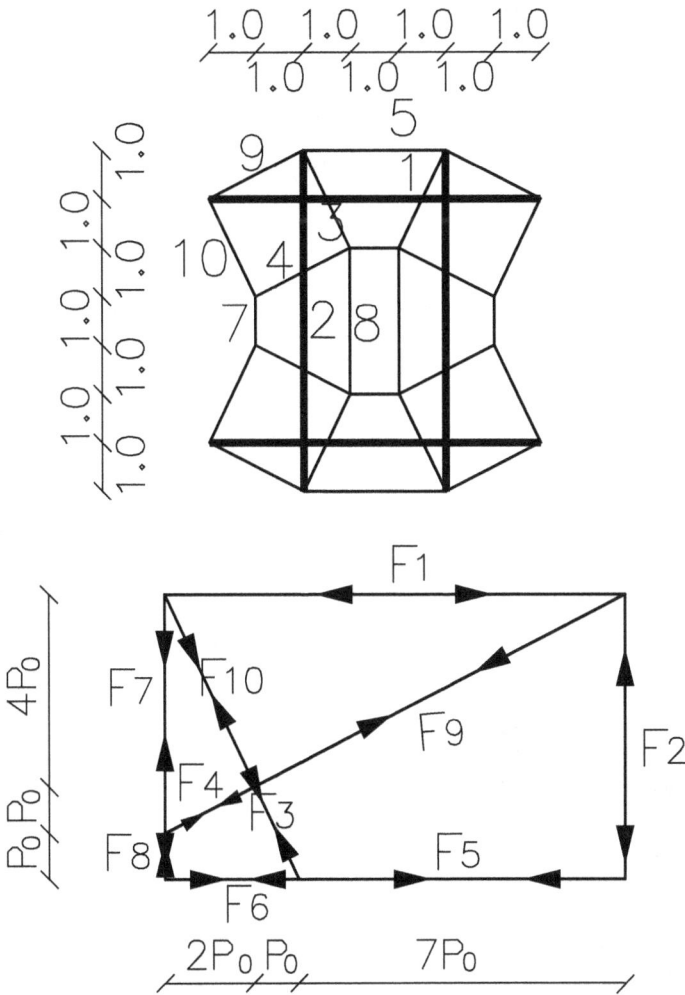

Figure 2.20 The basic sect of the tensegrity net shown in Figure 2.10.

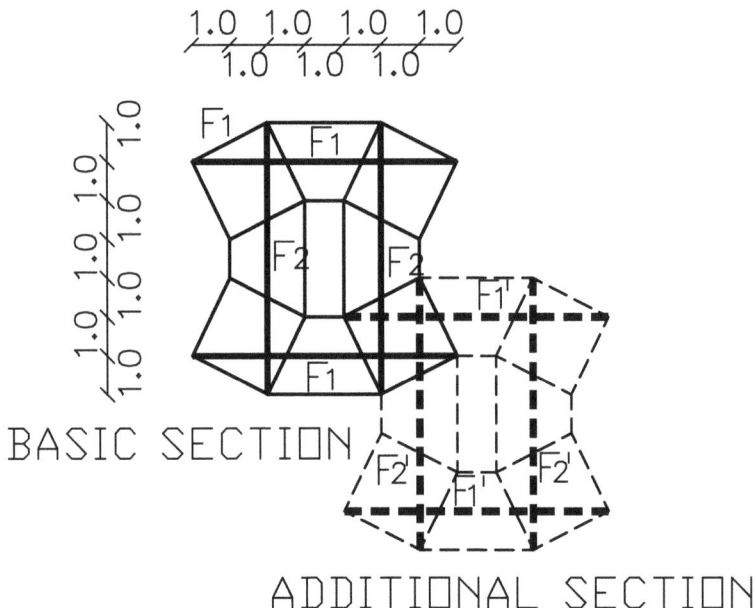

Figure 2.21 Additional section to the tensegrity net shown in Figure 2.21.

It can be seen that there are two degrees of freedom in prestressing of this section F_1 and F_2. It is possible to determine F_1 and F_2 independently. The forces induced to the members of the basic section where F_1 and F_2 are assumed to be $10P_0$ and $6P_0$, respectively, is shown in Figure 2.20. Also, this tensegrity net can be prestressed in one go or section by section as described earlier. The additional section is shown in Figure 2.21.

Also, in this case, the forces F_1' and F_2' of the additional section apply "fitted load" to the first previous section as shown by the force diagram in Figure 2.11.

In the case of the tensegrity net shown in Figure 2.13, there are two basic sections as shown in Figure 2.22.

It can be seen that by inducing $F_1{}^A$, the forces in the basic section A can be determined. By inducing $F_1{}^B$, the forces in basic section B can be determined.

The net can be prestressed in one go by the required forces $F_1{}^A$ and $F_1{}^B$ to every section that form the net. It can also be prestressed section by section by adding sections to each other as shown in Figure 2.23.

Also, in this case, the forces applied by prestressing a new section to the sections prestressed already are "fitted load" as shown in Figure 2.14. These forces change the forces in that section but not its configuration. The designer should determine the prestressing forces at each stage to achieve a desired result.

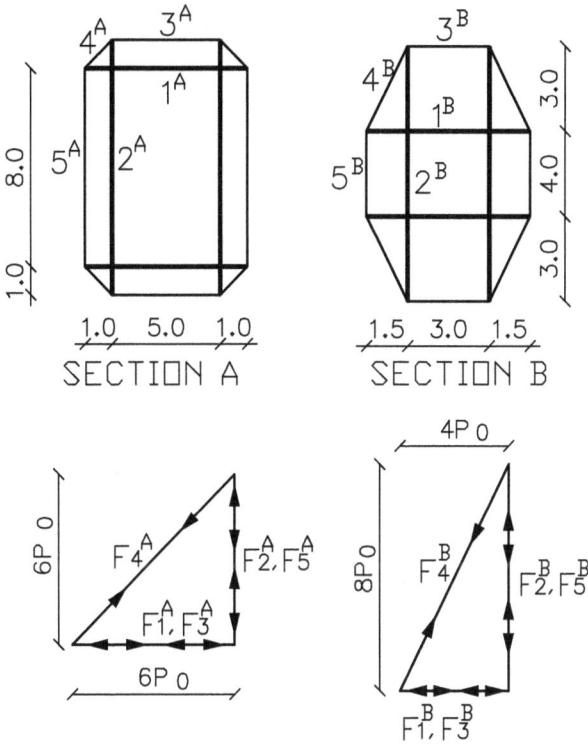

Figure 2.22 Two basic sects of the tensegrity net shown in Figure 2.13.

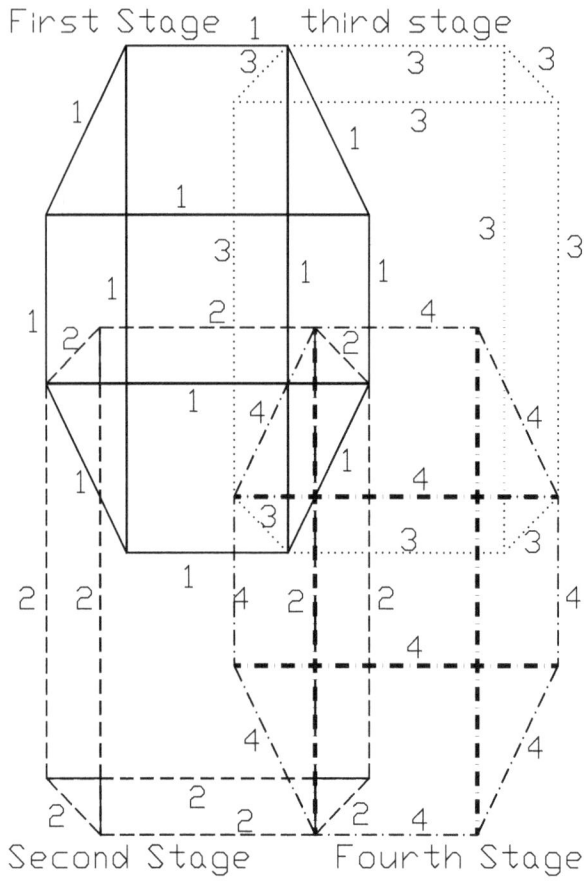

Figure 2.23 First four sections in the prestressing of the net shown in Figure 2.22.

Chapter 3

Tensegrity chains

Following the configuration of the tensegrity nets shown in Section 2.1, it is possible to construct tensegrity chains.

Examples of tensegrity chains similar to net F shown in Figures 2.10 and 2.12 are shown in Figure 3.1.

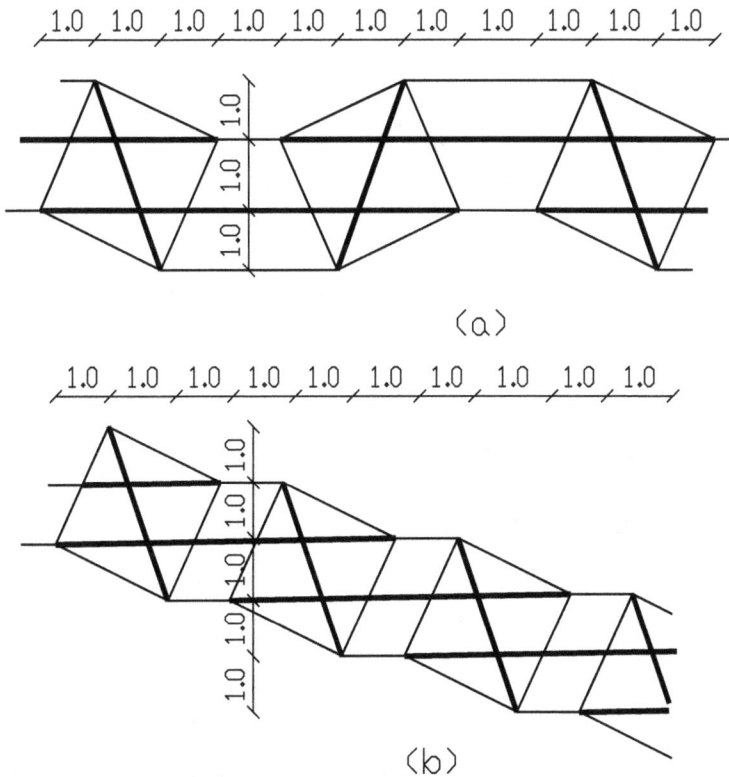

Figure 3.1 Examples of tensegrity chains.

DOI: 10.1201/9781003370093-4

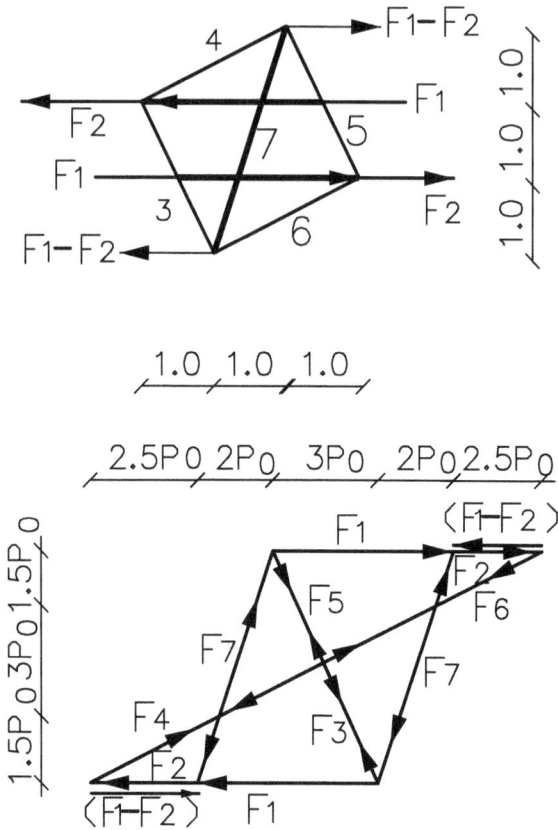

Figure 3.2 Forces induced by prestressing of the chain shown in Figure 3.1.

These tensegrity chains are prestressable. Prestressing forces and overall equilibrium of the chain under prestressing are shown in Figure 3.2.

Overall equilibrium implies:

$$3.1 \quad (F_1 + F_2) = 3(F_1 - F_2); \quad F_2 = F_1 / 2; \quad F_1 = 5P_0; \quad F_2 = 2.5P_0$$

The fact that equilibrium is maintained under the prestressing forces and compression is induced into the bars and tension to the cables indicates that this tensegrity chain is an appropriately prestressed tensegrity chain. It is a feasible tensegrity structure and can be constructed in this configuration.

The tensegrity net in Figure 2.13 leads to the tensegrity chains shown in Figure 3.3.

Prestressing forces of the tensegrity chain shown in Figure 3.3a are shown in Figure 3.4.

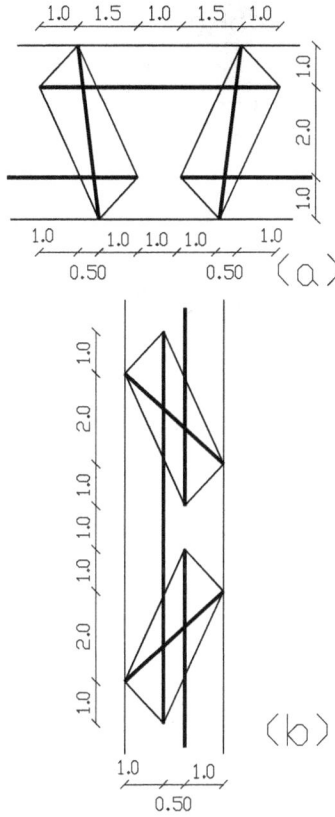

Figure 3.3 Tensegrity chains following the tensegrity nets shown in Figures 2.13.

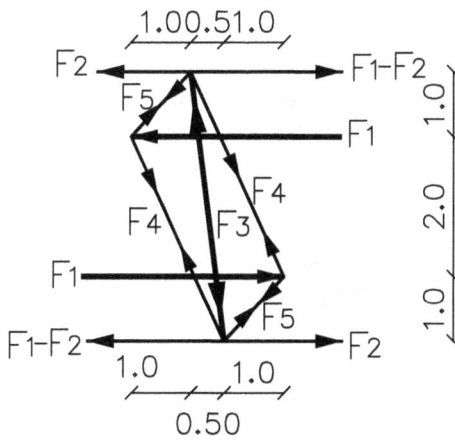

Figure 3.4 Prestressing forces of the tensegrity chain shown in Figure 3.3a.

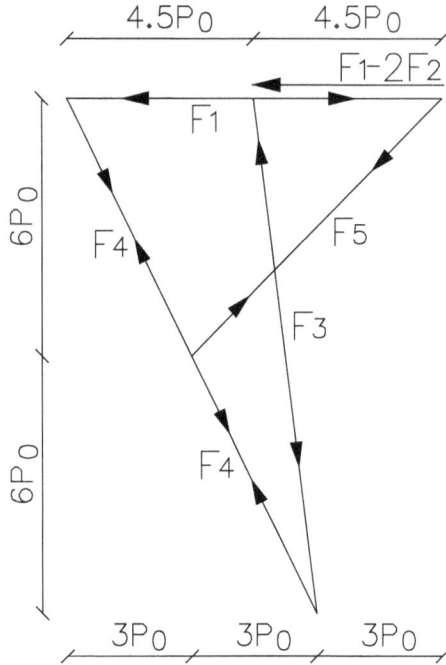

Figure 3.5 Force diagram of the tensegrity chain shown in Figure 3.3a.

Overall equilibrium implies:

$$3.2 \quad 2F_1 = 4\left(F_1 - 2F_2\right) F_2 = F_1 / 4; \quad F_1 = 9P_0; \quad F_2 = 9P_0 / 4$$

In the case where it is assumed that $F_1 = 9P_0$, the force diagram of the pre-stressing forces takes the form shown in Figure 3.5.

The fact that equilibrium is maintained under prestressing forces and compression is induced to the bars and tension to the cables indicates that this tensegrity chain is an appropriately prestressed tensegrity chain. It is a feasible tensegrity structure and can be constructed in the configuration shown in Figure 3.3a.

The prestressing forces applied to the tensegrity chain shown in Figure 3.3b are shown in Figure 3.6.

Overall equilibrium implies:

$$3.3 \quad F_1 = 5(F_1 - 2F_2) F_2 = 0.4F_1$$

In the case where it is assumed that $F_1 = 15P_0$, the force diagram of prestressing the tensegrity net presented in Figure 3.3b takes the form shown in Figure 3.7.

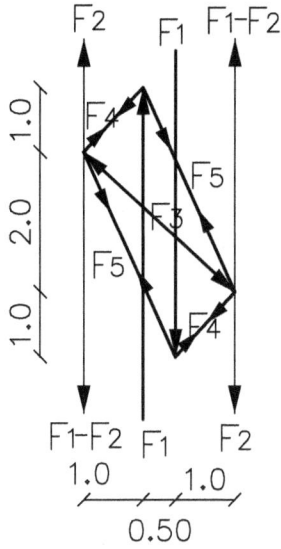

Figure 3.6 Prestressing forces applied to the tensegrity chain shown in Figure 3.3b.

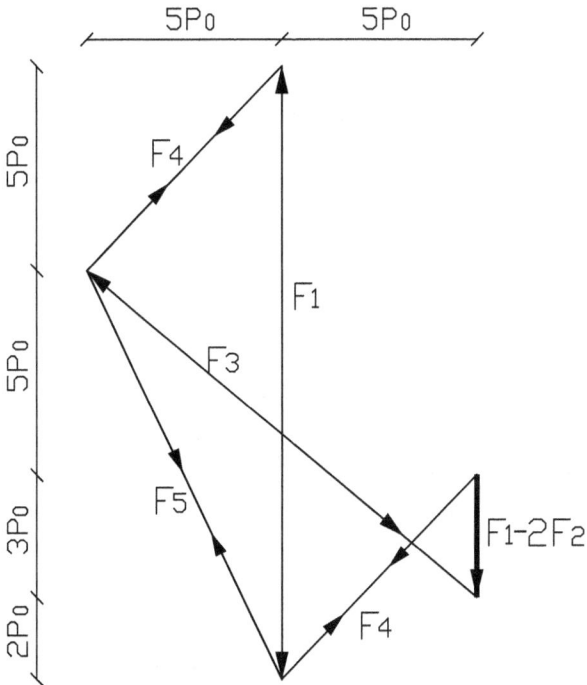

Figure 3.7 Prestressing force diagram of the tensegrity chain shown in Figure 3.3b.

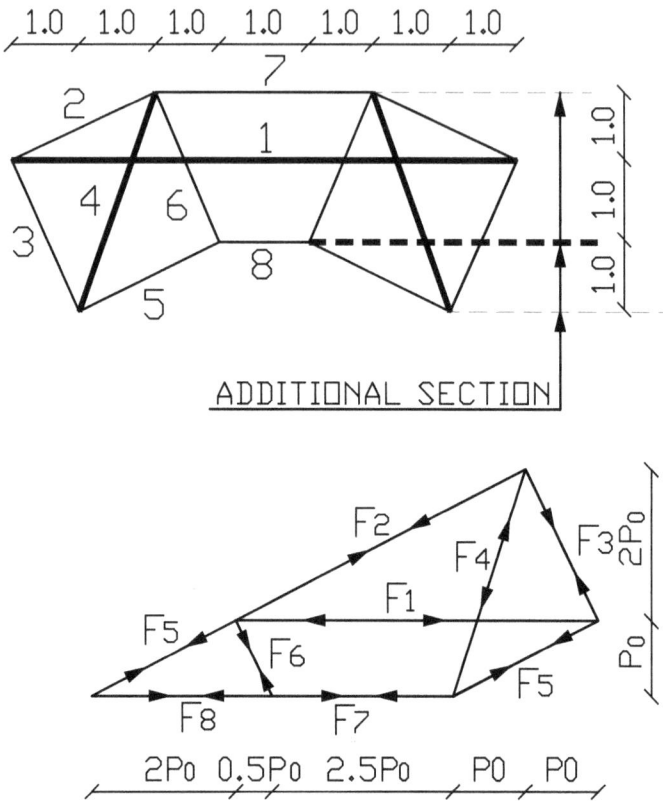

Figure 3.8 Basic section of the tensegrity chain shown in Figure 3.1a.

The fact that equilibrium is maintained under the prestressing forces and compression is induced to the bars and tension to the cables indicates that this tensegrity chain is appropriately prestressable and so it is a feasible tensegrity structure and can be constructed in this configuration.

Prestressing tensegrity chains can be carried out by using similar methods presented in Section 2.2 for prestressing tensegrity nets. The basic section of prestressing the tensegrity chain shown in Figure 3.1a is shown in Figure 3.8 as well as the prestressing forces.

The additional section is shown in Figure 3.8. The forces applied to the basic section when the additional section is prestressed are "fitted load", as shown by the force diagram in Figure 3.8.

The basic section of the chain shown in Figure 3.1b is shown in Figure 3.9.

The additional section is shown in Figure 3.9. Also, in this case, prestressing the additional section applies "fitted load" to the basic section as shown in the force diagram in Figure 3.9.

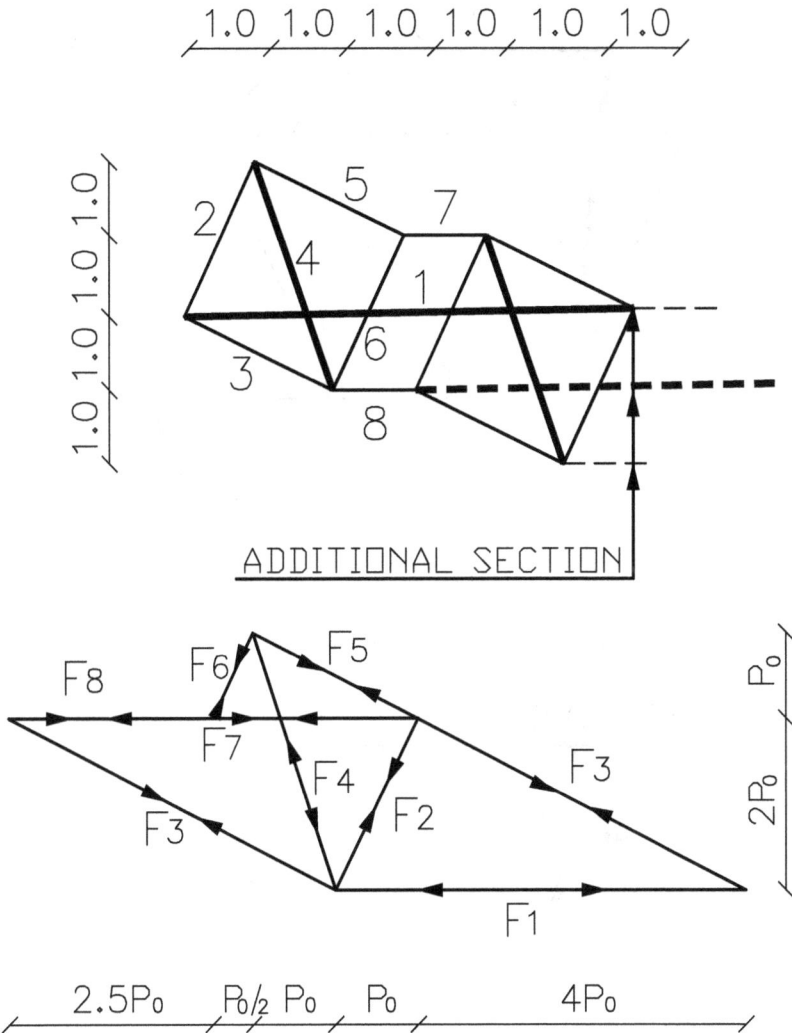

Figure 3.9 Basic section of the tensegrity chain shown in Figure 3.1b.

The basic sections of the tensegrity chains shown in Figure 3.3a and 3.3b are shown in Figures 3.10 and 3.11, respectively.

It can be seen that in these cases, the additional sections apply "fitted load" to the appropriate basic sections.

Obviously, it is possible to prestress all sections of the tensegrity chains simultaneously in one go. It is also possible to prestress the tensegrity chain step by step. In this case, the forces applied by prestressing an additional section to the part of the tensegrity chain that was prestressed already are

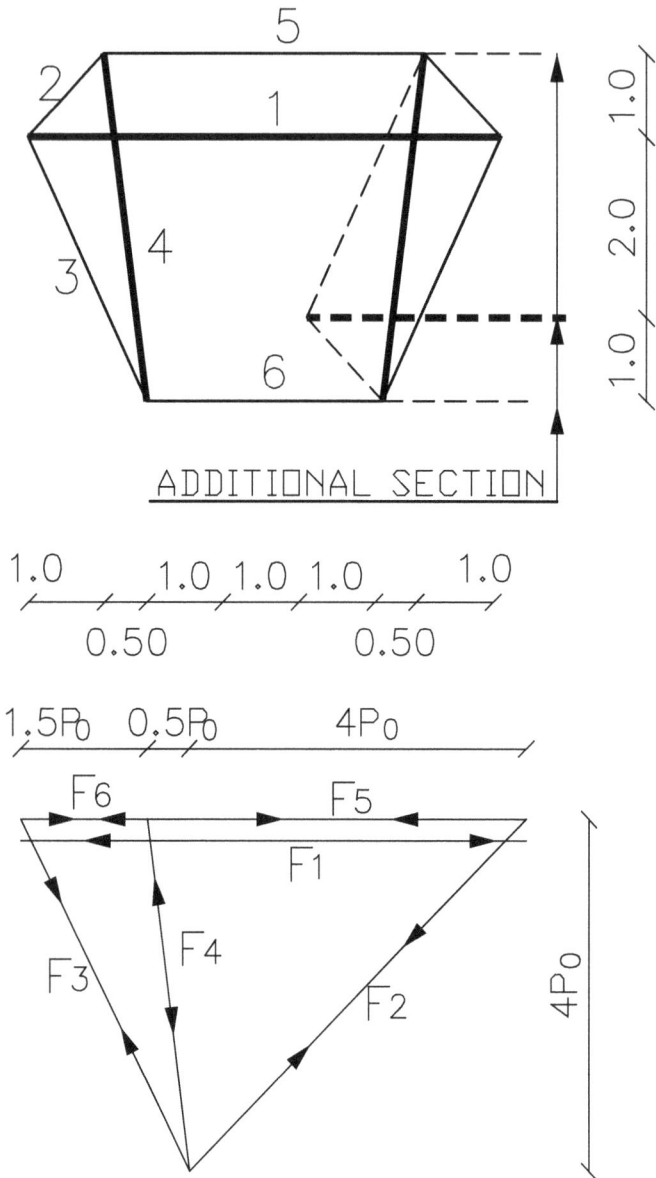

Figure 3.10 Basic section of the tensegrity chain shown in Figure 3.3a.

"fitted load". The change of the configuration of the tensegrity chain that was prestressed already is marginal and the change of the prestressing forces in the part of the tensegrity chain that was prestressed already can be established by following the methods presented in Vilnay (1990).

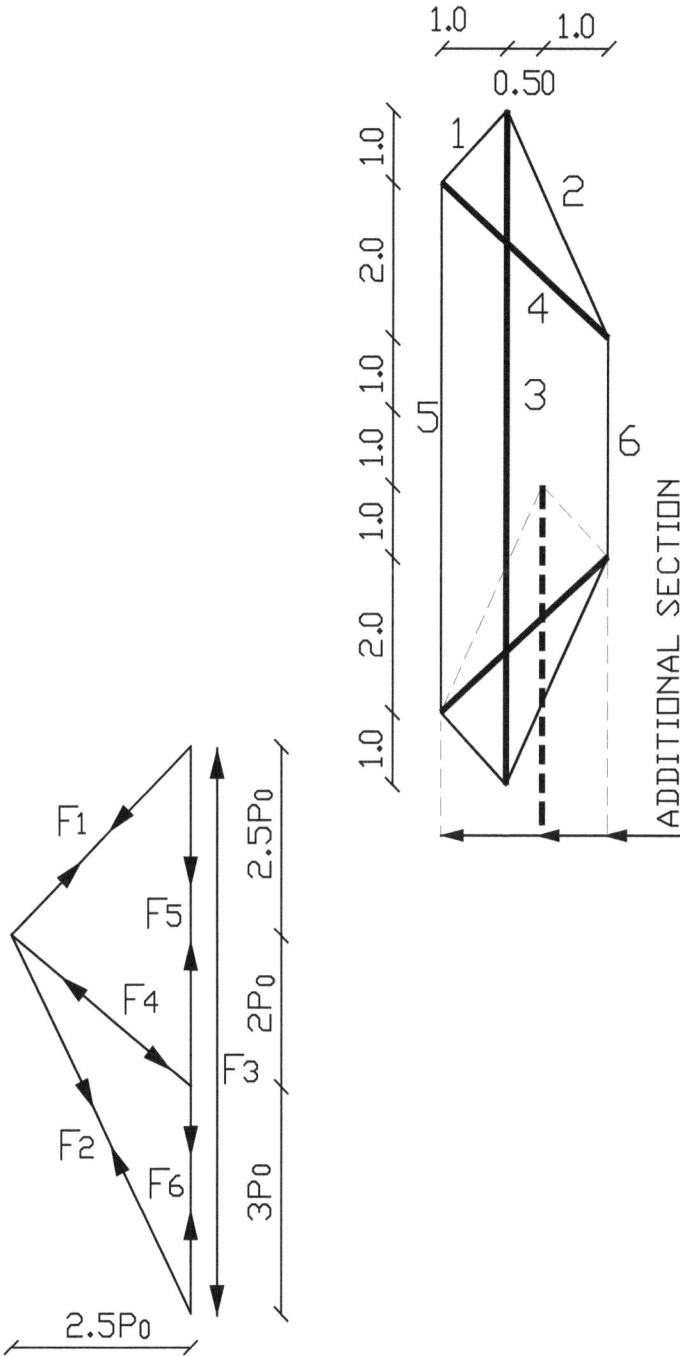

Figure 3.11 Basic section of the tensegrity chain shown in Figure 3.3b.

Chapter 4

Tensegrity plane arches

The simple square kite shown in Figure 1.2 can take the shape of an arch, as shown in Figure 4.1.

This arch is composed of a continuous tensile cable net and compression bars which are connected to the cable net only and not to each other. Since all bars are attached to the foundations, this arch cannot be considered to be a genuine tensegrity structure.

A more elaborate tensegrity arch based on the kite in Figure 1.10 is shown in Figure 4.2.

This tensegrity arch is composed of a continuous cable net and compression bars, the bars are connected to the cable net only and not to each other. One bar is not attached to the foundations. So this may be considered to be a genuine tensegrity arch.

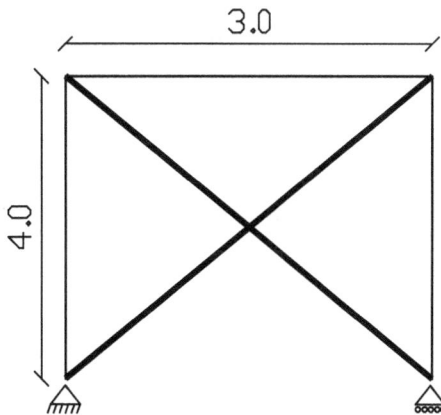

Figure 4.1 Arch constructed following the simple square kite shown in Figure 1.2.

DOI: 10.1201/9781003370093-5

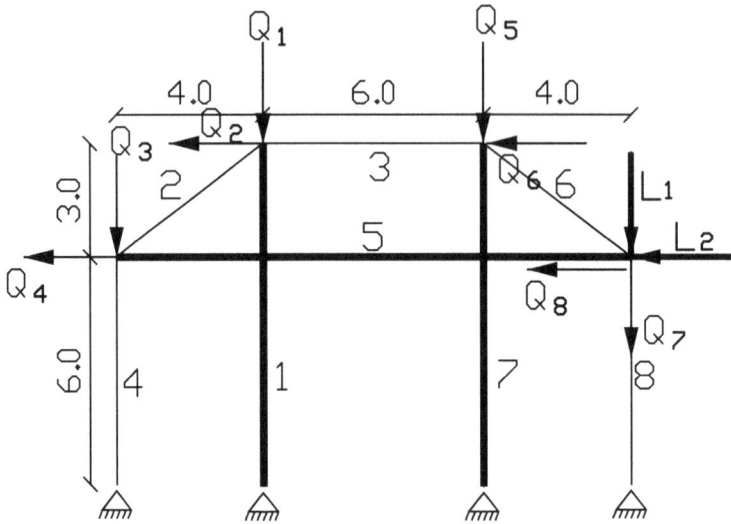

Figure 4.2 Tensegrity arch.

The equilibrium matrix **A** 8 x 8 takes the following form:

$$
4.1 \quad \mathbf{A} =
\begin{vmatrix}
-1 & -3/5 & 0 & 0 & 0 & 0 & 0 & 0 \\
0 & -4/5 & 1 & 0 & 0 & 0 & 0 & 0 \\
0 & 3/5 & 0 & -1 & 0 & 0 & 0 & 0 \\
0 & 4/5 & 0 & 0 & 1 & 0 & 0 & 0 \\
0 & 0 & 0 & 0 & 0 & -3/5 & -1 & 0 \\
0 & 0 & -1 & 0 & 0 & 4/5 & 0 & 0 \\
0 & 0 & 0 & 0 & 0 & 3/5 & 0 & -1 \\
0 & 0 & 0 & 0 & -1 & -4/5 & 0 & 0
\end{vmatrix}
\begin{matrix}
L1 \\ L2 \\ L3 \\ L4 \\ L5 \\ L6 \\ L7 \\ L8
\end{matrix}
$$

Here, Li indicates the appropriate line of matrix A.
It can be seen that the lines of matrix A are not independent.

$$
4.2 \quad L2 + L4 + L6 + L8 = 0
$$

Equation 4.2 indicates that the rank of matrix **A** is only 7, which implies that this tensegrity arch is an infinitesimal mechanism, prestressable and has one degree of freedom in determining the prestressing forces. It can be pre-stressed by lengthening one of the bars or by shortening one of the cables. In this way, the force in one element of the tensegrity arch is established; by using the equilibrium Equation 1.1, the forces induced to all other elements of the tensegrity arch can be predicted. The magnitude of the prestressing force is up to the designer.

By using Equation 1.4 and assuming that bar 1 is prestressed to $6P_0$, the prestressing forces of this tensegrity arch take the following form:

$$4.3 \quad \mathbf{P} = P_0 \begin{vmatrix} -6 \\ 10 \\ 8 \\ 6 \\ -8 \\ 10 \\ -10 \\ 6 \end{vmatrix}$$

By using Equation 4.2, the "fitted load" takes the following form:

$$4.4 \quad \mathbf{Q} = \begin{vmatrix} Q1 \\ Q2 \\ Q3 \\ Q4 \\ Q5 \\ Q6 \\ Q7 \\ -Q2 - Q4 - Q6 \end{vmatrix}$$

Vector $\mathbf{Q}_1$ considering load $\mathbf{L1}$ shown in Figure 4.2 takes the following form:

$$4.5 \quad \mathbf{Q}_1 = \begin{vmatrix} 0 \\ 0 \\ 0 \\ 0 \\ 0 \\ 0 \\ \mathbf{L1} \\ 0 \end{vmatrix}$$

Vector $\mathbf{Q}_1$ considering load **L2** shown in Figure 4.2 takes the following form:

$$4.6 \quad \mathbf{Q}_2 = \begin{vmatrix} 0 \\ 0 \\ 0 \\ 0 \\ 0 \\ 0 \\ 0 \\ \mathbf{L2} \end{vmatrix}$$

Because load $\mathbf{Q}_1$ satisfies Equation 4.4, it is "fitted load". The tensegrity arch can sustain it in the prestressed configuration. The change in the forces in the tensegrity arch elements due to the load **L1** can be found by using the methods presented in Vilnay (1990).

Because load $\mathbf{Q}_2$ does not satisfy Equation 4.4, it is not "fitted load". The tensegrity arch cannot sustain it in its prestressable configuration. When loaded by such a load, the configuration of the tensegrity arch changes and the so-called large infinitesimal displacements take place. The magnitude of the displacements can be predicted by following Vilnay (1990).

It is possible to determine the prestressing forces induced by prestressing by using the appropriate force diagram.

The force diagram of the tensegrity arch, where it is assumed that cable 1 is prestressed to 6 P_0, $F_1 = 6P_0$, is shown in Figure 4.3.

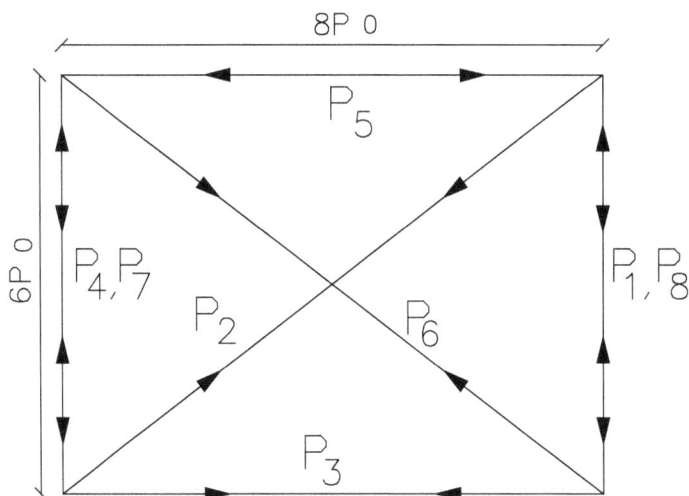

Figure 4.3 Force diagram of prestressing the tensegrity arch shown in Figure 4.2.

The prestressing forces shown in Figure 4.3 are similar to those given by Equation 4.3.

The cases of "fitted load" and not "fitted load" can also be studied by using the equilibrium method presented in Chapter 1.

In the case where **L1** shown in Figure 4.2 is $2P_0$ and the tensegrity arch is prestressed by inducing $6P_0$ to cable 1 so that $F_1 = 6P_0$, the force diagram takes the form shown in Figure 4.4.

It can be seen that the tensegrity arch can be in equilibrium considering **L1**. This is an indication that the load given by Equation 4.5 is "fitted load". This result is consistent with Equation 4.4.

In the case where the tensegrity arch shown in Figure 4.2 is loaded by **L2** = $2P_0$ and cable 1 is prestressed to $6P_0$ so that $F_1 = 6P_0$, the force diagram takes the form shown in Figure 4.5.

It can be seen that the tensegrity arch cannot be in equilibrium in the prestressed configuration. Cable 8 is not in the "right" direction. This fact implies that **L2** is not "fitted load". This result is also consistent with the condition given by Equation 4.4.

In the case where the tensegrity arch is prestressed and then loaded with **L1**, the load changes the forces in the tensegrity arch members. In tensegrity structure, the equilibrium equation implies that it is also possible to adjust the prestressing force at any step to a required level and so to control the forces in the tensegrity arch members where it is loaded by **L1**. So the tensegrity arch can be prestressed to gain its configuration and the prestressing can

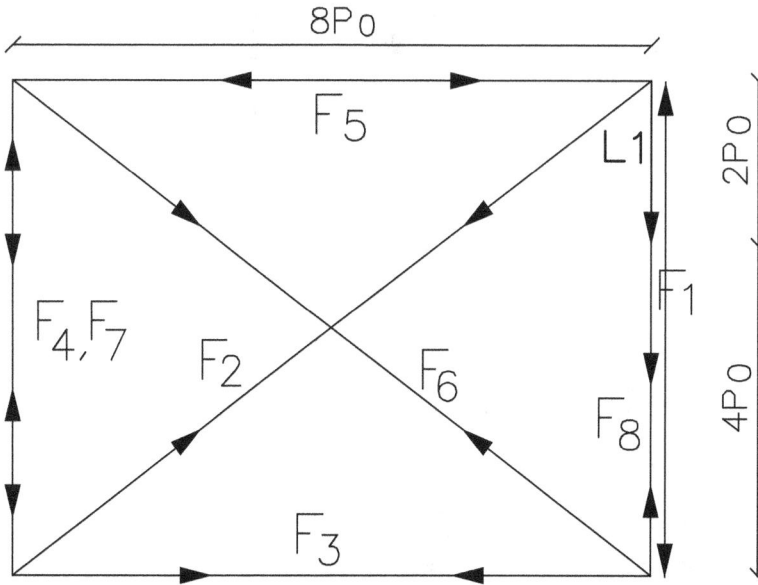

Figure 4.4 Force diagram of the tensegrity arch shown in Figure 4.2 loaded by **LI** = $2P_0$.

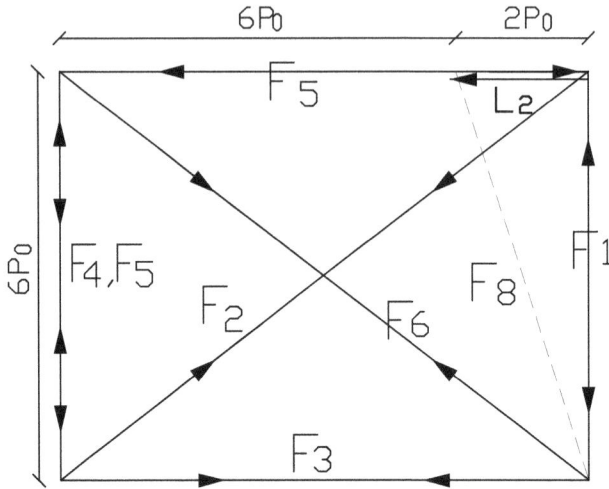

Figure 4.5 Force diagram of the tensegrity arch shown in Figure 4.2 loaded by **L2** = $2P_0$.

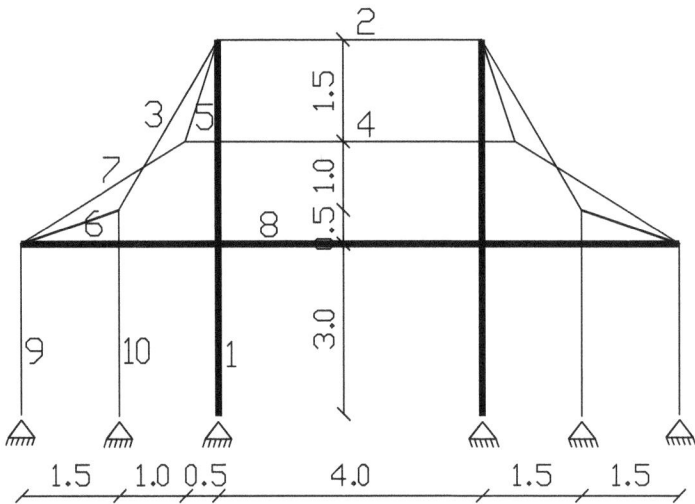

Figure 4.6 Interesting tensegrity arch.

be maintained simultaneously with the load **L1**. The prestressing can also be adjusted to its original level at a later stage. In this case, the forces in the tensegrity arch members are shown in Figure 4.4.

The magnitude of prestressing should be to the level required to keep all cables under tension considering the applied load **L1**.

Tensegrity arch with two prestressing degrees of freedom is shown in Figure 4.6.

This arch can be seen as composed of two arches (Figure 4.7).

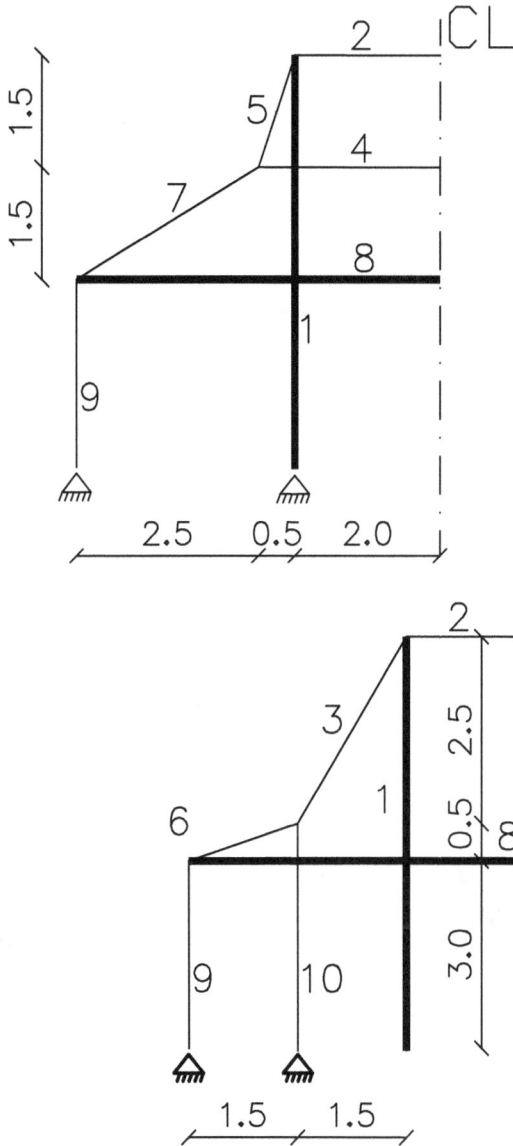

Figure 4.7 Two tensegrity arches that compose the tensegrity arch shown in Figure 4.6.

Figure 4.7 indicates that there are two degrees of freedom in prestressing this tensegrity arch. It can be prestressed by prestressing one member of 4, 5 or 7 and one member of 3, 6 or 10 simultaneously. It can also be prestressed step by step. For example, at the first step, one cable 4, 5 or 7 and in the second stage one cable 3, 6 or 10. The prestressing forces of the second stage apply "fitted load" to the tensegrity arch prestressed at the first stage. Thus, the tensegrity arch that was prestressed already can sustain this load in its configuration. The change in the forces of the first stage of prestressing due to the second stage of prestressing can be found in Vilnay (1990).

A tensegrity arch with some cables parallel to the bars is shown in Figure 4.8.

The tensegrity arch is prestressed by applying force equal to $3P_0$ to cable 1, the force diagram is shown in Figure 4.8. It indicates that in this case, the compression forces in the bars, P_1 and P_8, are of the same magnitude as the tension in cables P_4 and P_7, respectively:

$$4.7 \quad \begin{aligned} P_1 &= -P_2 \\ P_5 &= -P_4 \\ P_8 &= -P_7 \end{aligned}$$

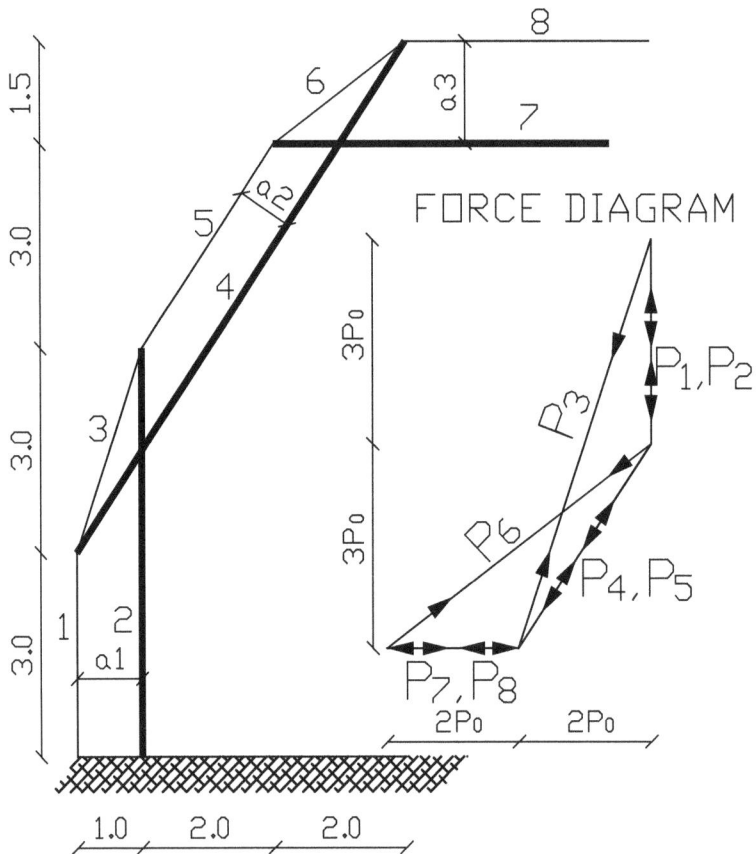

Figure 4.8 Tensegrity arch with some cables parallel to the bars.

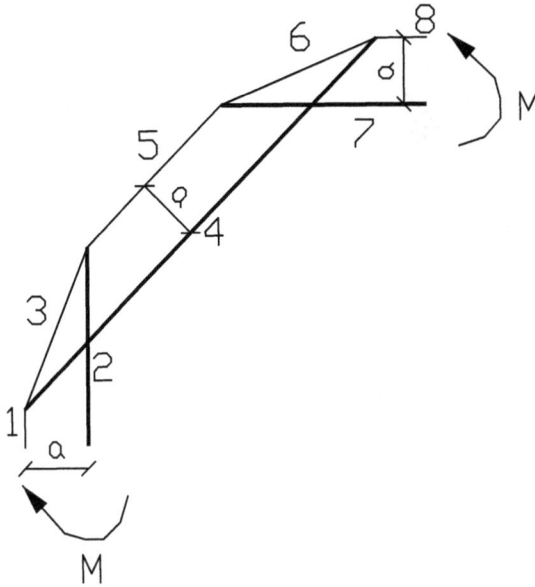

Figure 4.9 Tensegrity arch.

The case where the spacing between the bars and the appropriate parallel cables is the same is shown in Figure 4.9.

In this case, to maintain equilibrium of the moment M, shown in Figure 4.9, throughout the tensegrity arch, the forces induced to the tensegrity arch bars and the appropriate parallel cables are equal.

$$4.8 \quad P_1 = P_5 = P_8 = -P_2 = -P_4 = -P_7$$

The forces induced to the tensegrity arch bars and the appropriate parallel cables are the same throughout the tensegrity arch.

Not all proposed tensegrity arches are feasible; for example, the tensegrity arch shown in Figure 4.10 is not a feasible tensegrity arch.

From the force diagram, it can be seen that to maintain equilibrium, cable 8 cannot be horizontal. Thus, this proposed tensegrity arch is not a feasible tensegrity arch and cannot be constructed in this configuration.

In the design of a symmetrical tensegrity arch, the designer should be aware of the following points:

- The cable at the line of symmetry should be horizontal.
- The type of foundation: there are two extreme types of foundation, as shown in Figure 4.11.

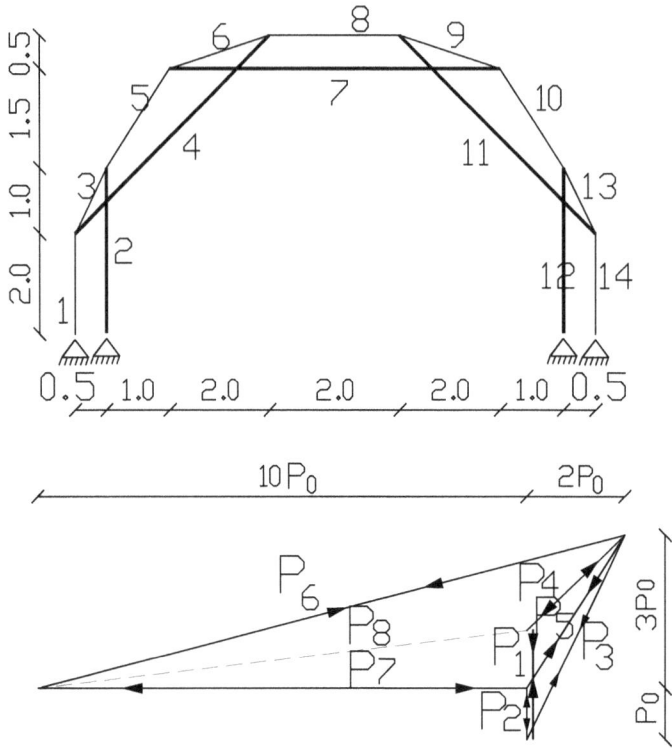

Figure 4.10 Unfeasible tensegrity arch.

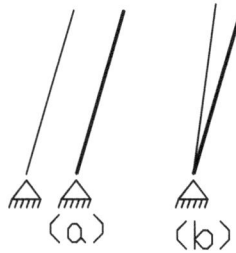

Figure 4.11 Two types of foundations of tensegrity arch.

Foundation type (Figure 4.11a) has two units: one, a foundation to the cable and another one to the bar, the cable is parallel to the associated bar. Foundation type (Figure 4.11b) is one foundation to the cable and to the bar. In the case of foundation type (Figure 4.11b), the arch configuration should satisfy the requirements shown in Figure 4.12. To achieve

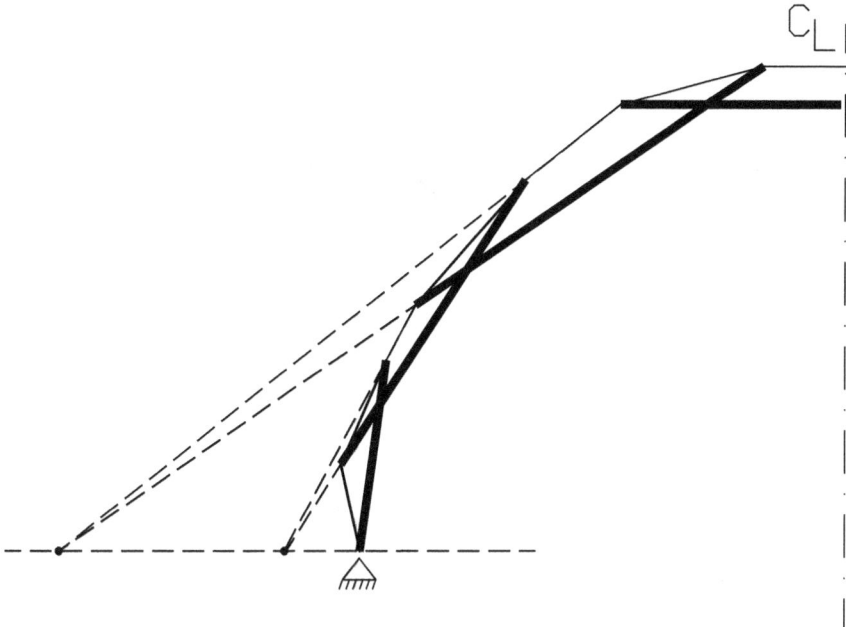

Figure 4.12 Conditions that the configuration of a tensegrity arch with a single foundation should satisfy.

equilibrium considering the prestressing forces, the interaction of each bar line of action and the line of action of the cable associated with it should be at the foundation level.

In the case of a type (Figure 4.11a) foundation, the cables should be parallel to the associated bars throughout the tensegrity arch as shown in Figure 4.13.

In this case, by selecting the magnitude of a_1, a_2 and a_3, the designer can control the forces in the cables and the bars as shown in Figure 4.8.

An important type of tensegrity arches is a tensegrity arch with a configuration based on the shape of a suspension cable.

The most common suspension cable loaded by equal loads at equal spacing is shown in Figure 4.14.

The forces induced to the cable elements can be found by using the force diagram shown in Figure 4.14.

By fitting bars in the proper locations, the suspension cable takes the configuration of a tensegrity arch. The tensegrity arch based on the suspension cable shown in Figure 4.14 takes the form shown in Figure 4.15.

The tensegrity arch shown in Figure 4.15 is a feasible tensegrity arch because bars a, b and c are parallel to cables 1, 3 and 5, respectively.

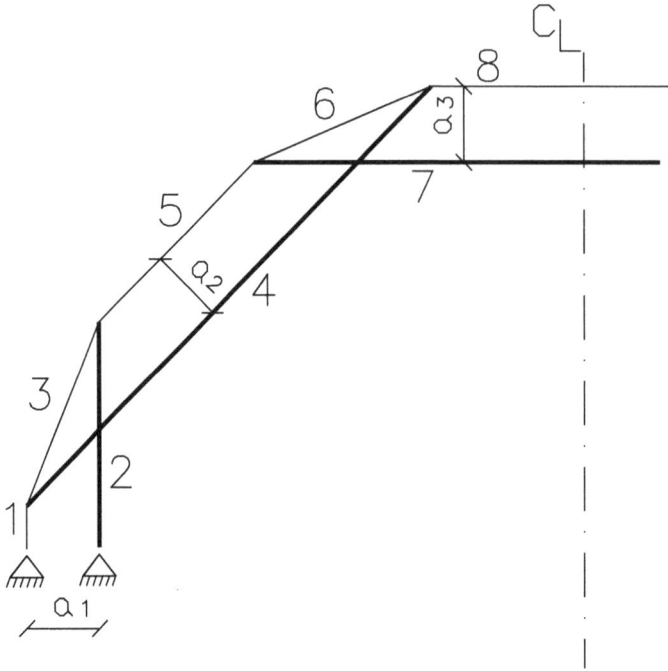

Figure 4.13 Configuration of a tensegrity arch with foundation type (Figure 4.11a).

The forces induced to the tensegrity arch members by prestressing are shown in Figure 4.16.

It can be seen that

$$
\begin{aligned}
P_x^4 &= P_x^2 = 6P_0 \\
P_x^5 &= P_x^3 = P_x^1 = 3P_0 \\
P_x^c &= P_x^a = P_x^b = -3P_0 \\
P_z^4 &= P_z^3 = 2P_0 \\
P_z^a &= -4P_0 \\
P_z^1 &= 4P_0 \\
P_z^2 &= 6P_0 \\
P_z^b &= -2P_0
\end{aligned}
\qquad 4.9
$$

Whether loading case $L1 = L4 = P_0$ is "fitted load" can be investigated by using the "equilibrium method" presented in Chapter 1. The force diagram

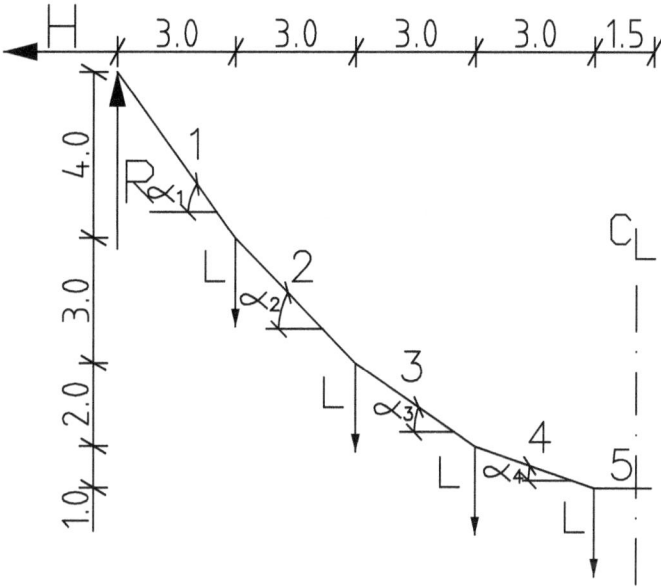

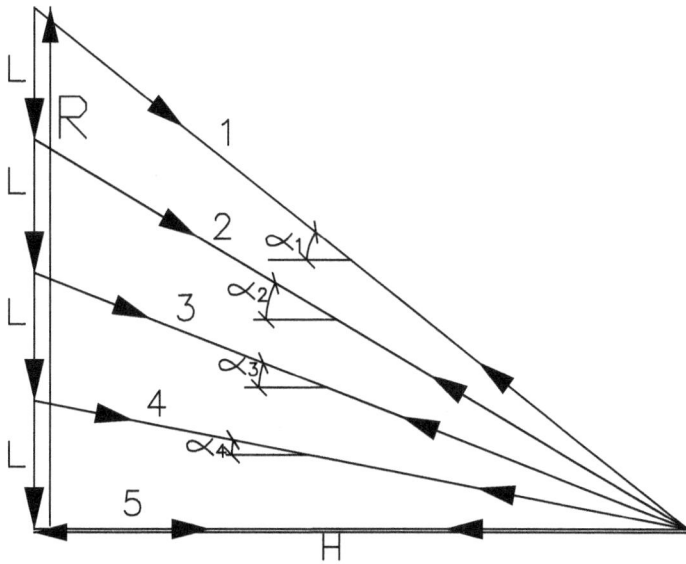

FORCE DIAGRAM

Figure 4.14 A suspension cable.

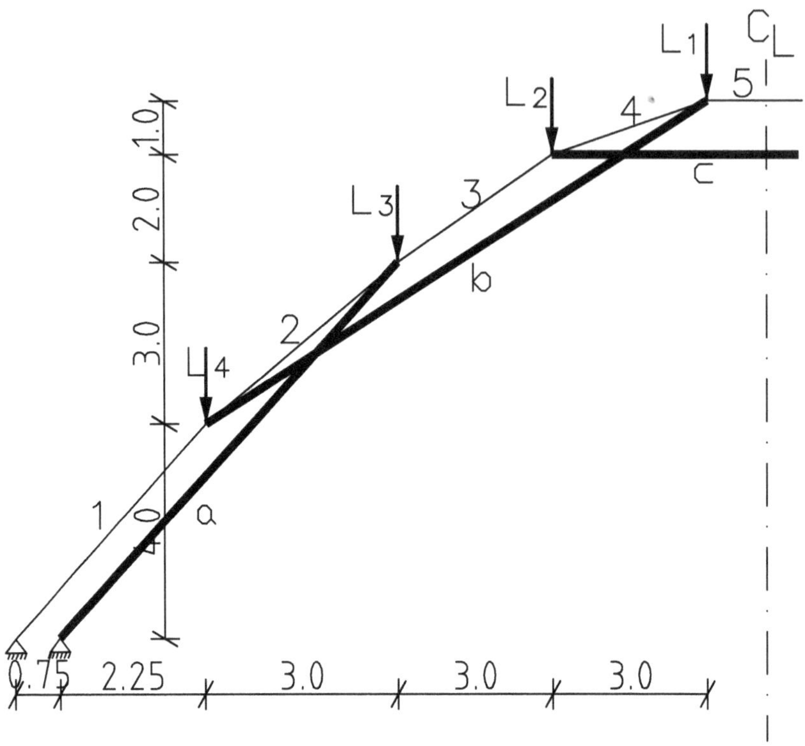

Figure 4.15 Tensegrity arch based on the suspension cable shown in Figure 4.14.

considering $L1 = L4 = P_0$ and prestressing force of $6\sqrt{2}\ P_0$ in cable 2 is shown in Figure 4.17a. The fact that it is a consistent force diagram indicates that $L1 = L4$ is "fitted load".

In the case where the tensegrity arch shown in Figure 4.15 is loaded by $L2$ and $L3$ where $L2 = L3 = P_0$ ($L1 = L4 = 0$) and it is simultaneously prestressed and the force in cable 2 is $6\sqrt{2}\ P_0$, the forces in the tensegrity arch elements are shown in Figure 4.17b. Also, in this case, the fact that the force diagram is consistent indicates that $L2 = L3 = P_0$ is "fitted load".

The case where the tensegrity arch shown in Figure 4.15 is loaded by $L1 = L2 = L3 = L4 = P_0$ and the tensegrity arch is simultaneously prestressed and the forces in cable 2 is $6\sqrt{2}\ P_0$, the forces in the tensegrity arch members are shown in Figure 4.18.

The consistent force diagram in Figure 4.18 indicates that the loading case $L1 = L2 = L3 = L4 = P_0$ is "fitted load". The tensegrity arch can sustain these loads in its prestressed configuration. Moreover, the designer can control the forces in the tensegrity arch members by adjusting the prestressing forces.

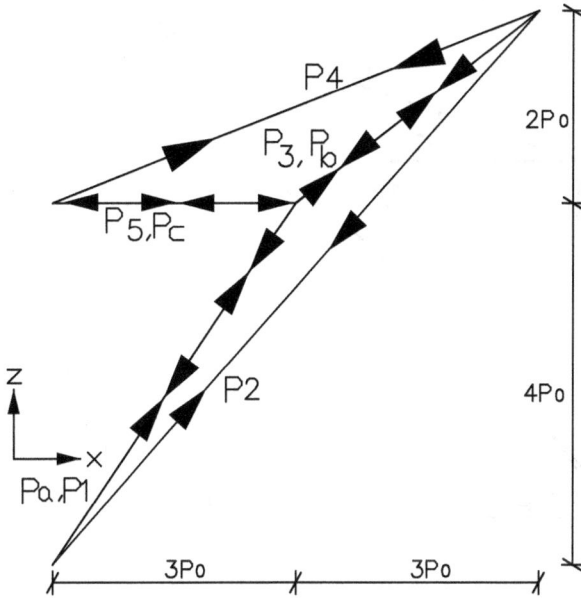

Figure 4.16 Prestressing forces of the tensegrity arch shown in Figure 4.15.

The fact that the loading cases in which $L1 = L4 = P_0$ and $L2 = L3 = 0$ and the loading case in which $L2 = L3 = P_0$ and $L1 = L4 = 0$ are "fitted loads" can also be studied by using the "mechanical method" presented in Chapter 1. By investigating the configuration of the tensegrity arch, it can be seen that when loaded with $L1 = L4 = P_0$, the load can actually be carried by members 1, b and 5, as shown in Figure 4.19a. The consistent force diagram in Figure 4.19a indicates a possible change of the forces in members 1, b and 5 due to the given load. It implies that the given load is "fitted load". The same is true in the case where the tensegrity arch is loaded by $L2 = L3 = P_0$, the load can be carried by members a, 3 and c, as shown in Figure 4.19b. Thus, the "mechanical method" also confirms that these loads are "fitted loads".

It is interesting that only the load $L1 = L2 = L3 = L4 = P_0$ can be considered to be a kind of "fitted load" of the suspension cable, but in the case of a tensegrity arch built following this suspension cable, the loadings $L1 = L4 = P_0$ and $L2 = L3 = P_0$ are also "fitted loads".

The procedure of adapting a suspension cable to form a tensegrity arch is not always successful. For example, in the case of the suspension cable shown in Figure 4.20, because bar b is paralleled to cable 3 and bars a and c are designed to be parallel to cables 1 and 5, respectively, this tensegrity arch is prestressable and so it is a feasible tensegrity arch.

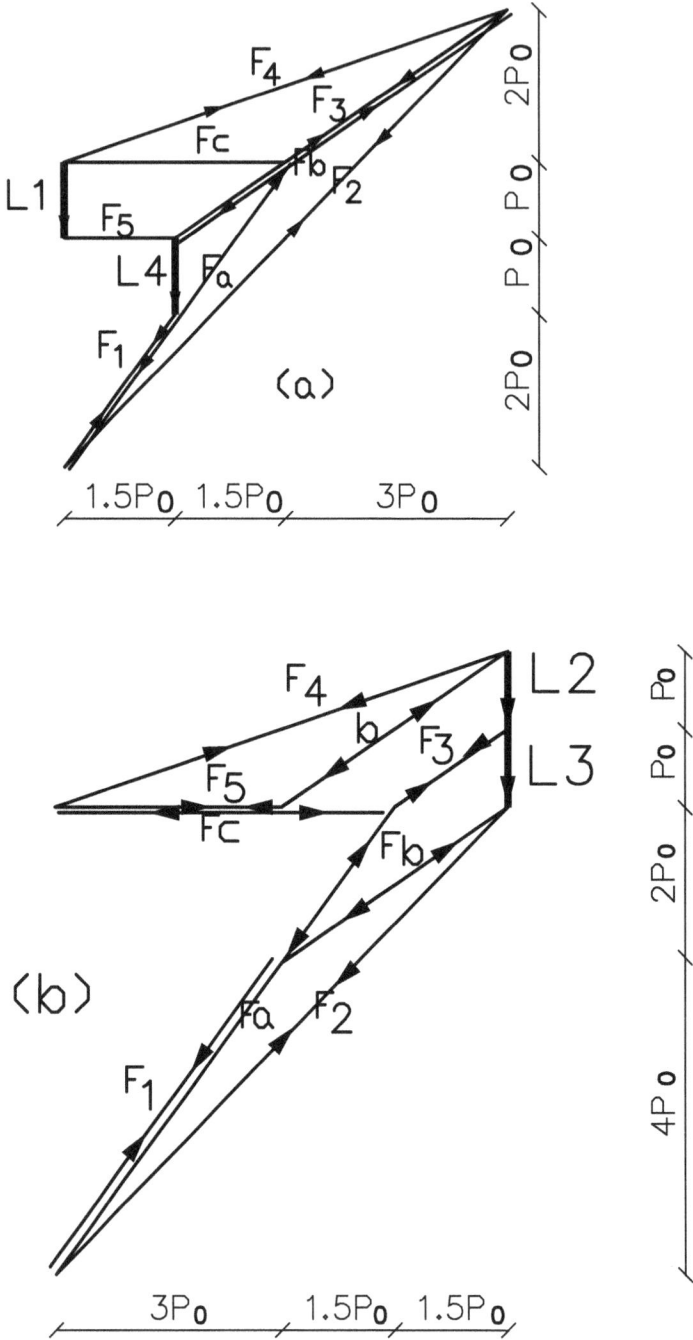

Figure 4.17 Force diagram due to **L1 = L4** and **L2 = L3**.

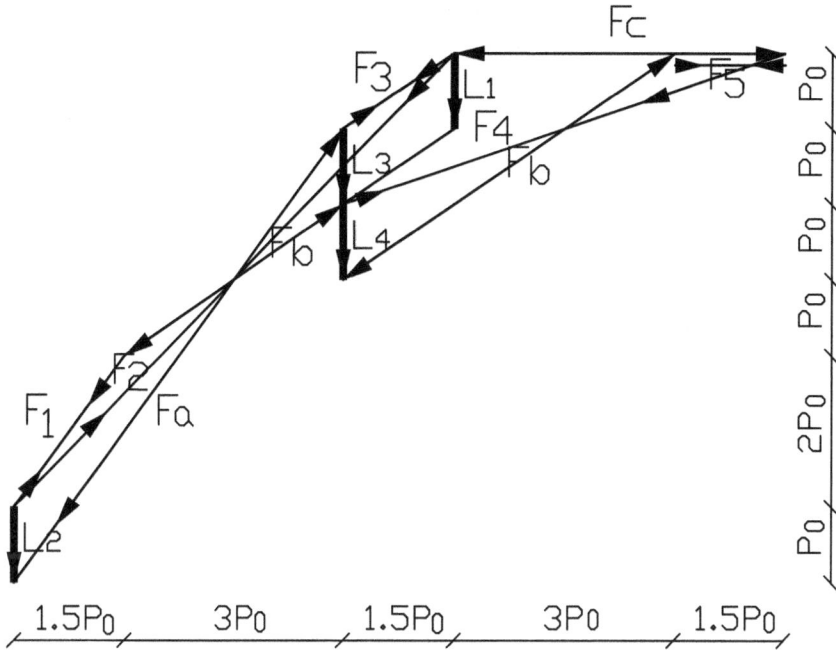

Figure 4.18 Forces induced into the tensegrity arch due to **L1** = **L2** = **L3** = **L4** = P_0.

In the case of the suspension cable shown in Figure 4.21, the obvious tensegrity arch formed by using this suspension cable is shown there. Because in this case the interaction of the lines of bar b and cable 3 shown in Figure 4.20 is not at the level of the supports A and B, this is not a prestressable tensegrity arch and so it is not a feasible tensegrity arch.

It is possible to use the force diagram of a suspension cable to determine the configuration of a tensegrity arch with a type (a) foundation shown in Figure 4.11. The designer is free to choose the preferable force diagram of the suspension cable. For example, the force diagram of a suspension cable loaded by eight equal loads shown in Figure 4.22.

The designer is free to locate the foundation F and nodes a and b in accordance with the inclination of cables 4 and 5 in the suspension cable force diagram. Locations of the foundation and node b determine bar 8. By considering the inclination of cable 3, point A can be fixed. Line Aa determines bar 7 inclination. The location of node c is up to the designer. The intersection of cable 2 and bare 7 determines node d. Nodes c and d determine cable 1 and bar 6.

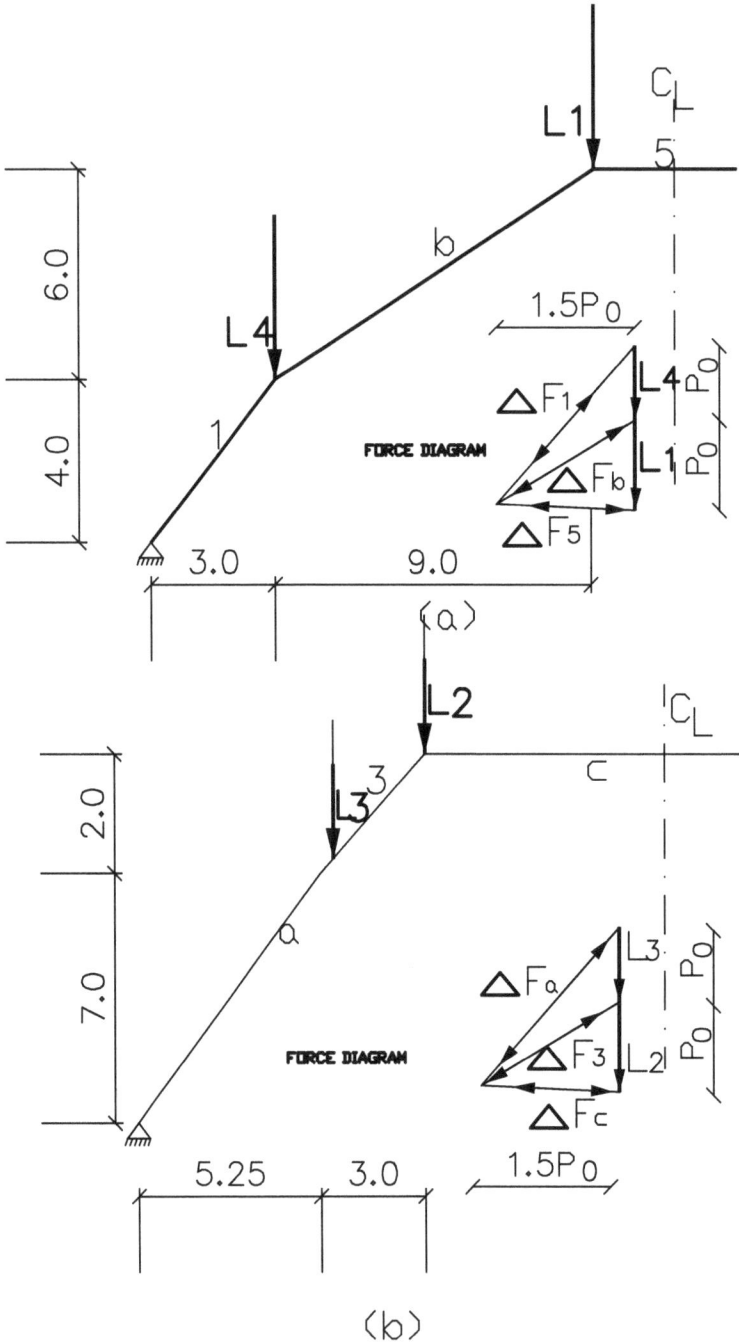

Figure 4.19 Structural units that supports **L1** = **L4** = P_0 and **L2** = **L3** = P_0.

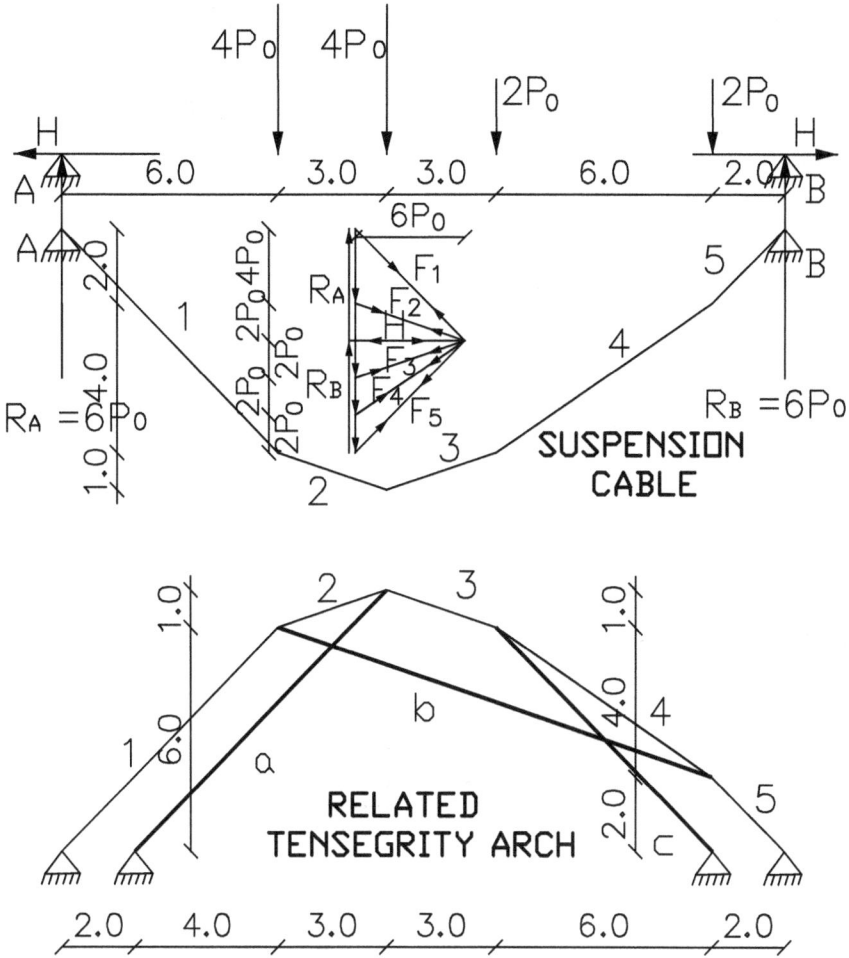

Figure 4.20 Feasible tensegrity arch constructed from a suspension cable.

The directions of bars 6, 7 and 8 in the suspension cable force diagram indicates that L_1 and L_2 are also "fitted load" of the tensegrity arch as well as P_1 and P_2.

The force diagram of the tensegrity arch where bar 8 is prestressed to $P_0\sqrt{337}$ is shown in Figure 4.23.

The fact that it is a consistent force diagram confirms that it is a feasible tensegrity arch and it can be constructed in the proposed configuration.

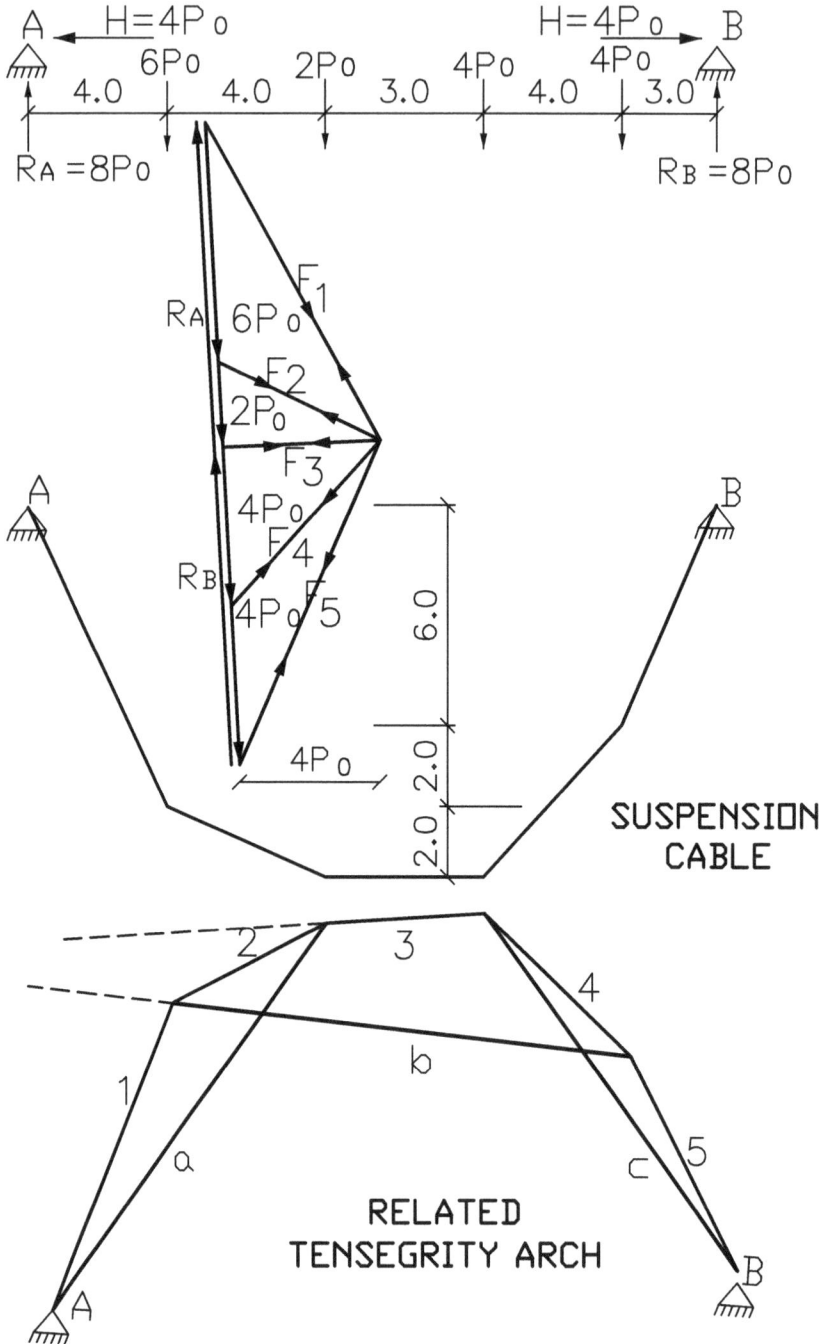

A ← H=4P₀

H=4P₀ → B

$A \leftarrow H=4P_0$

$6P_0$ $2P_0$ $4P_0$ $H=4P_0$ B

4.0   4.0   3.0   4.0   3.0

$R_A = 8P_0$

$R_B = 8P_0$

F_1
R_A $6P_0$
F_2
$2P_0$
F_3
$4P_0$
F_4
R_B
$4P_0$ F_5

$4P_0$

6.0

2.0 2.0

SUSPENSION
CABLE

2
3
4
b
1
a
c 5

RELATED
TENSEGRITY ARCH

A

B

Figure 4.21 Unsuccessful design of a feasible tensegrity arch from a suspension cable.

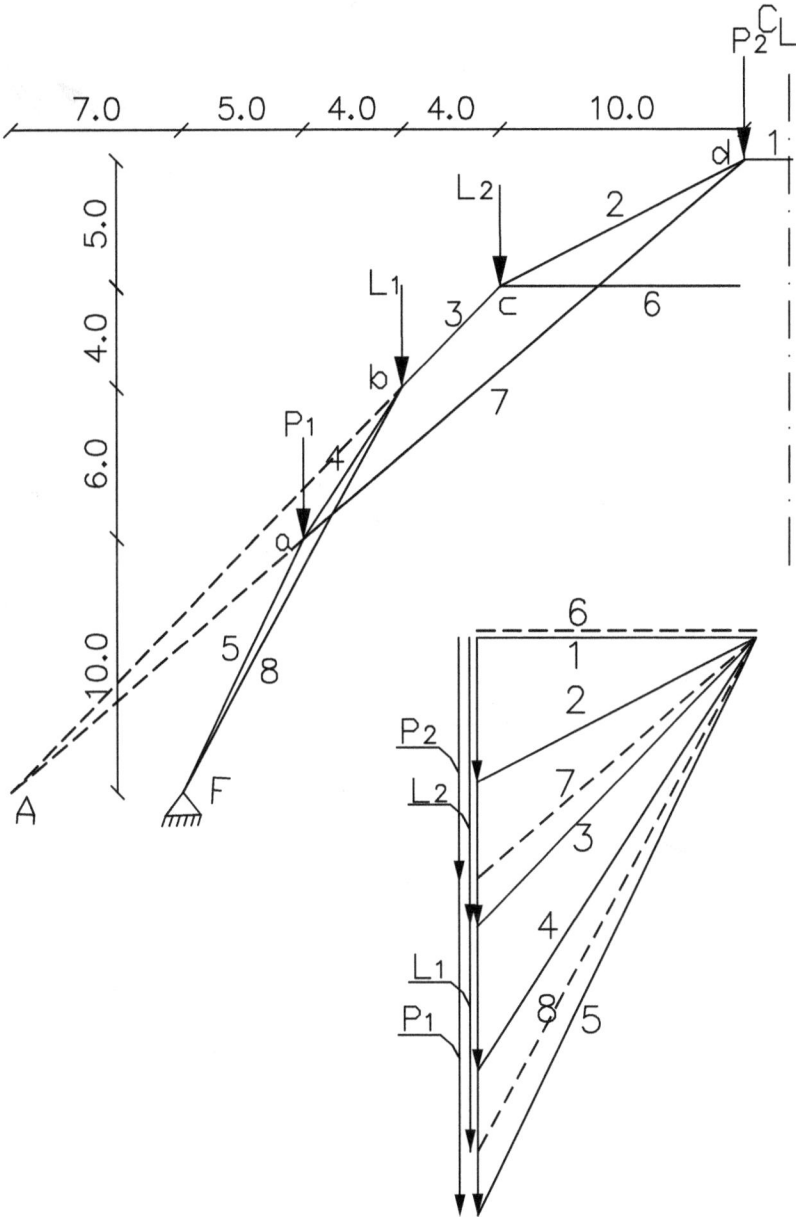

Figure 4.22 Tensegrity arch with type (a) foundation.

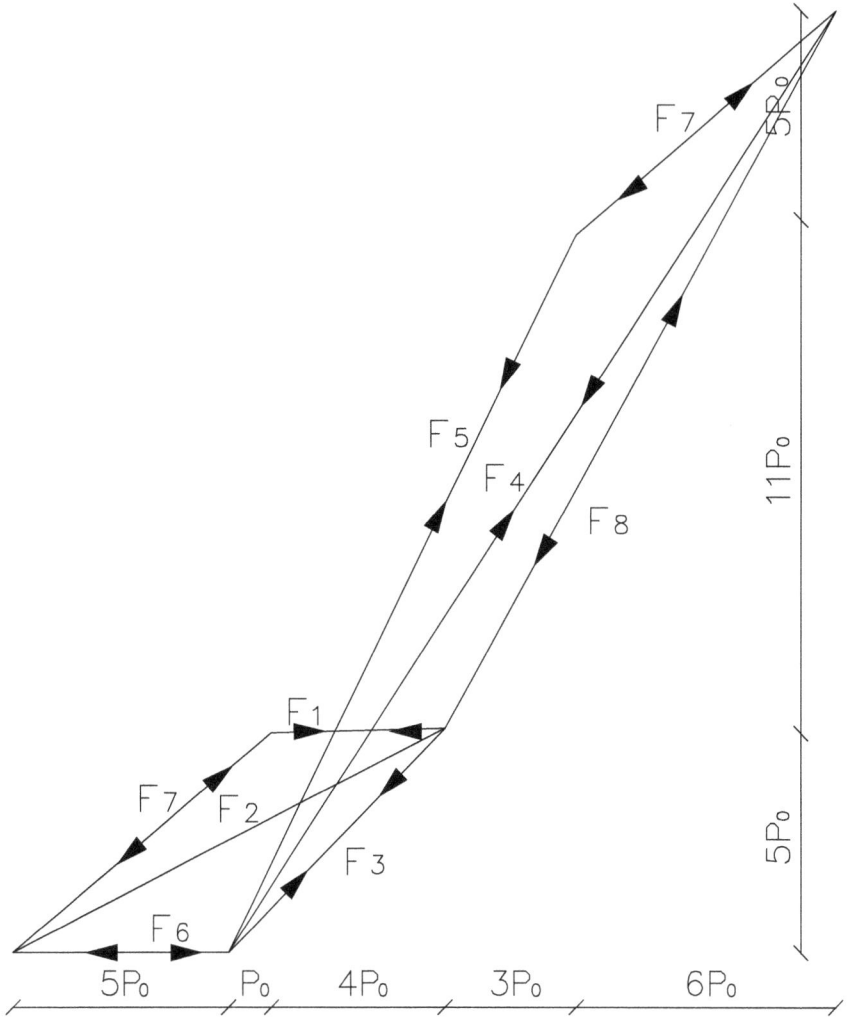

Figure 4.23 Force diagram of the tensegrity arch shown in Figure 4.22.

Chapter 5

Tensegrity rings

The tensegrity arch shown in Figure 4.2 can be extended to form a symmetrical tensegrity plane ring as shown in Figure 5.1.

This symmetrical ring can be prestressed, the prestressing forces are shown in Figure 5.2.

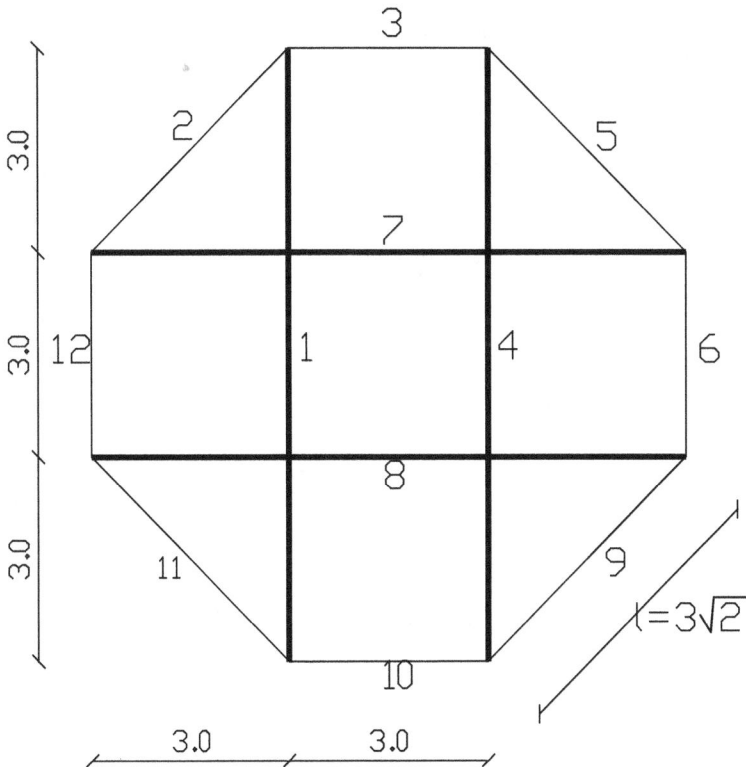

Figure 5.1 Symmetrical tensegrity plane ring.

DOI: 10.1201/9781003370093-6

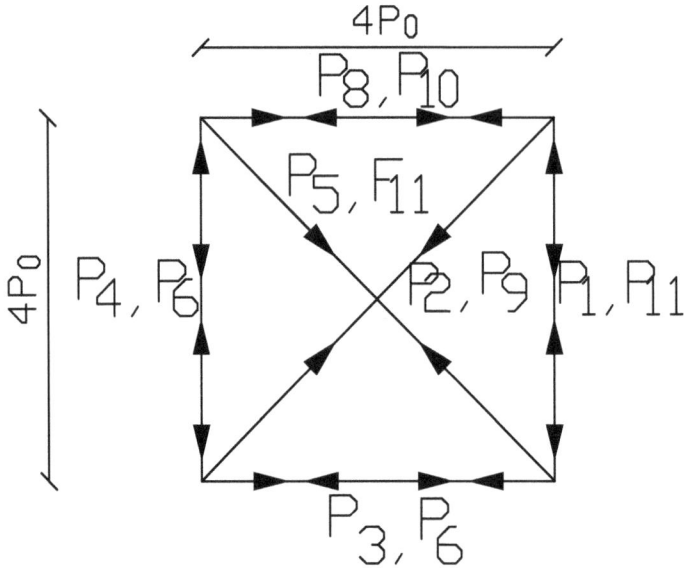

Figure 5.2 Force diagram of the prestressing forces of the tensegrity ring shown in
Figure 5.1.

It can be seen that in this symmetrical ring, the forces are the same in the identical members.

$$
\begin{aligned}
P_1 &= P_4 = P_7 = P_8 = -4P_0 \\
P_3 &= P_6 = P_{10} = P_{12} = 4P_0 \\
P_2 &= P_5 = P_9 = P_{11} = 4\sqrt{2}P_0
\end{aligned}
$$

5.1

Because the symmetrical tensegrity ring is appropriately prestressable, it is a feasible tensegrity ring.

It is possible to form a quasi-symmetrical ring as shown in Figure 5.3.

The prestressing force diagram of the quasi-symmetrical tensegrity ring is shown in Figure 5.4.

In this case,

$$5.2 \quad P_2 = P_5 = P_9 = P_{11} = \sqrt{5}P_0$$

and

$$5.3 \quad P_1 = P_4 = -2P_0 \quad P_6 = P_{12} = 2P_0 \quad P_7 = P_8 = -P_0 \quad P_3 = P_{10} = P_0$$

The tensegrity ring can take an odd shape as the one shown in Figure 5.5.

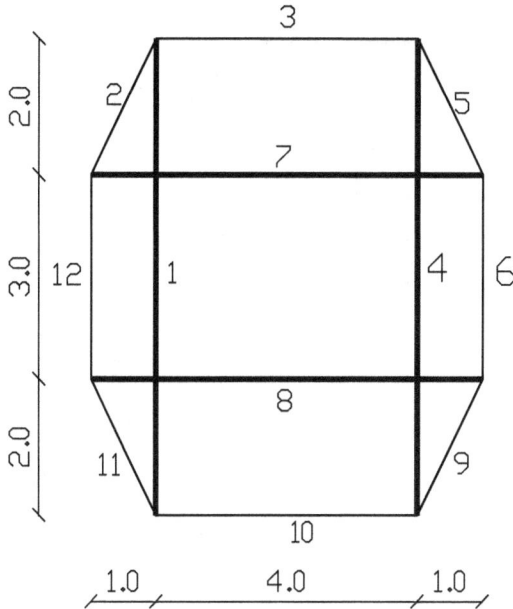

Figure 5.3 Quasi-symmetrical tensegrity ring.

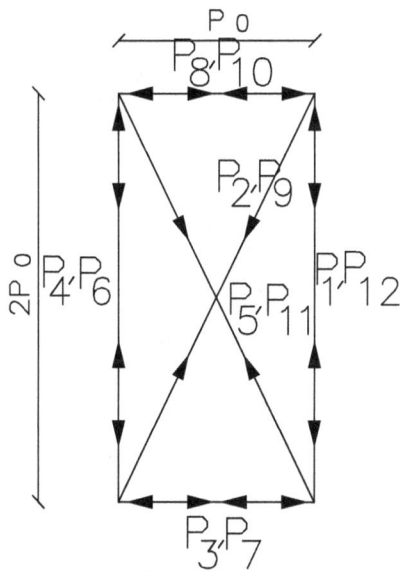

Figure 5.4 Force diagram of the ring shown in Figure 5.3.

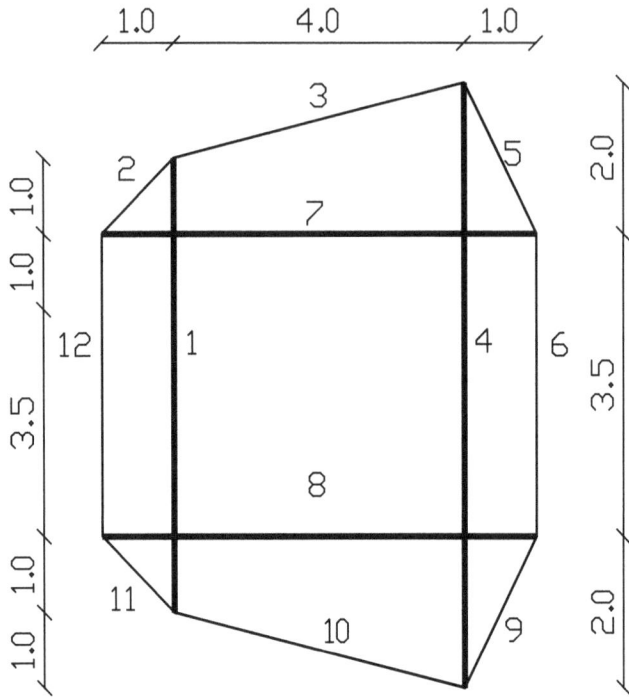

Figure 5.5 Odd shape tensegrity ring.

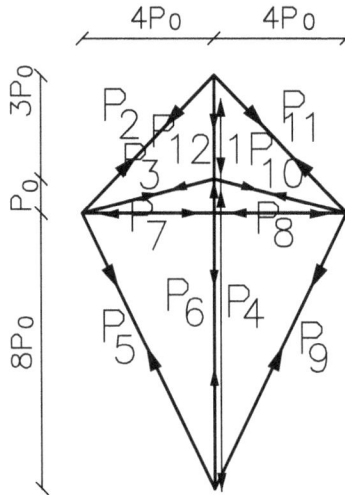

Figure 5.6 Force diagram of the prestressing forces of the odd shape tensegrity ring shown in Figure 5.5.

The prestressing force diagram of the tensegrity ring shown in Figure 5.5 is shown in Figure 5.6.

In this case,

$$5.4 \quad \begin{aligned}
P_1 &= -3P_0 \\
P_2 &= P_{11} = 4\sqrt{2}P_0 \\
P_3 &= P_{10} = \sqrt{17}P_0 \\
P_4 &= -9P_0 \\
P_5 &= P_9 = \sqrt{80}P_0 \\
P_6 &= 8P_0 \\
P_7 &= P_8 = -4P_0 \\
P_{12} &= 4P_0
\end{aligned}$$

The fact that the odd tensegrity ring shown in Figure 5.5 is appropriately prestressable implies that it is a feasible tensegrity ring.

Not all tensegrity rings are feasible; for example, the ring shown in Figure 5.7.

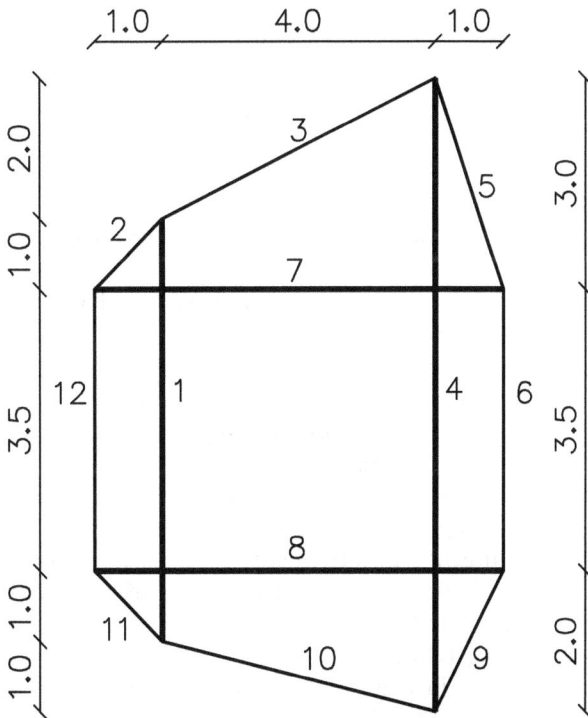

Figure 5.7 Proposed tensegrity ring.

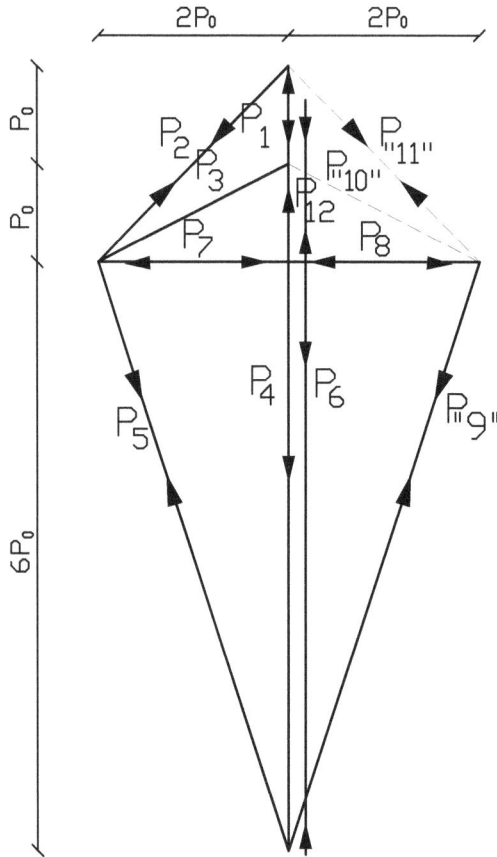

Figure 5.8 Prestressing force diagram of the proposed tensegrity ring shown in Figure 5.7.

The ring is prestressed by inducing a unit force P_0 to bar 1. The force diagram of the ring is shown in Figure 5.8.

It can be seen that the ring is not in equilibrium when prestressed. Cables 10 and 9 are not in the "right" direction to satisfy equilibrium. This fact indicates that this proposed tensegrity ring is not feasible and cannot be constructed in this configuration.

Tensegrity ring with six bars is shown in Figure 5.9.

The tensegrity ring is prestressed by inducing $3P_0$ to bar 1. The consistent force diagram in Figure 5.9 indicates that this is a feasible tensegrity ring and it can be constructed in the given configuration. A very similar symmetrical tensegrity ring is shown in Figure 5.9.

This tensegrity ring is prestressed by inducing $7P_0$ to bar 1. The prestressing force diagram in Figure 5.10 is not consistent as cable 11 is not in the "right" direction. So this tensegrity ring is not feasible and cannot

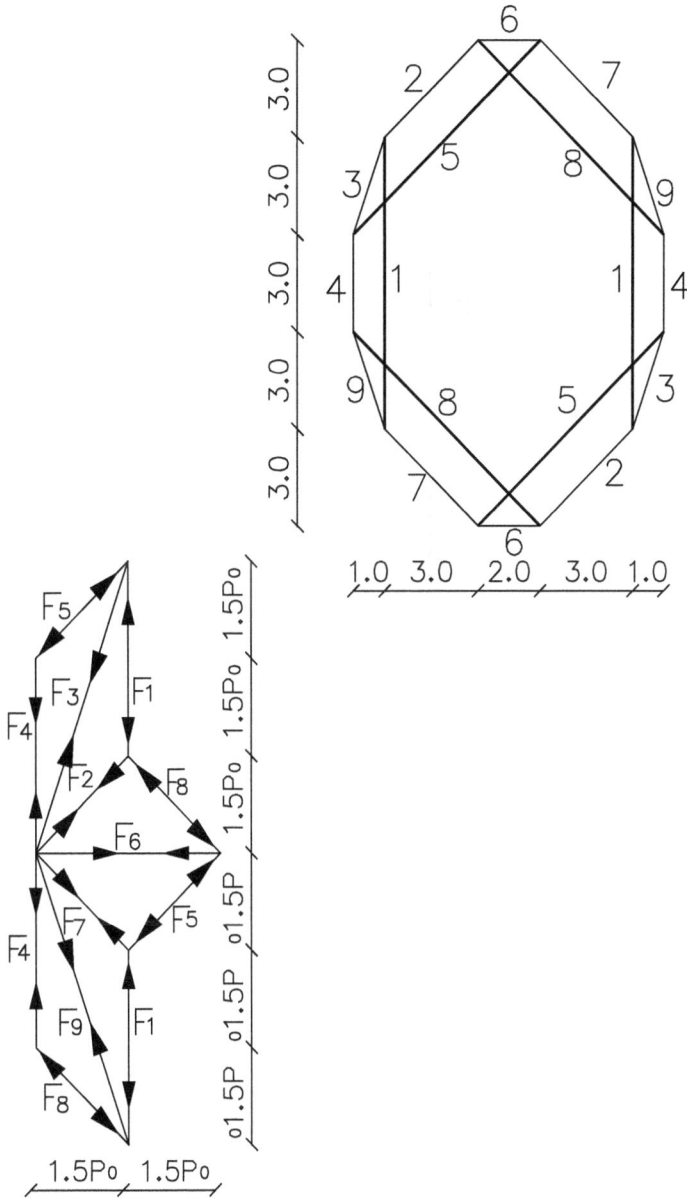

Figure 5.9 Tensegrity ring with six bars.

be constructed in this configuration. Comparing the symmetrical tensegrity rings in Figures 5.5 and 5.10 reveals that symmetry does not imply the feasibility of the tensegrity structure.

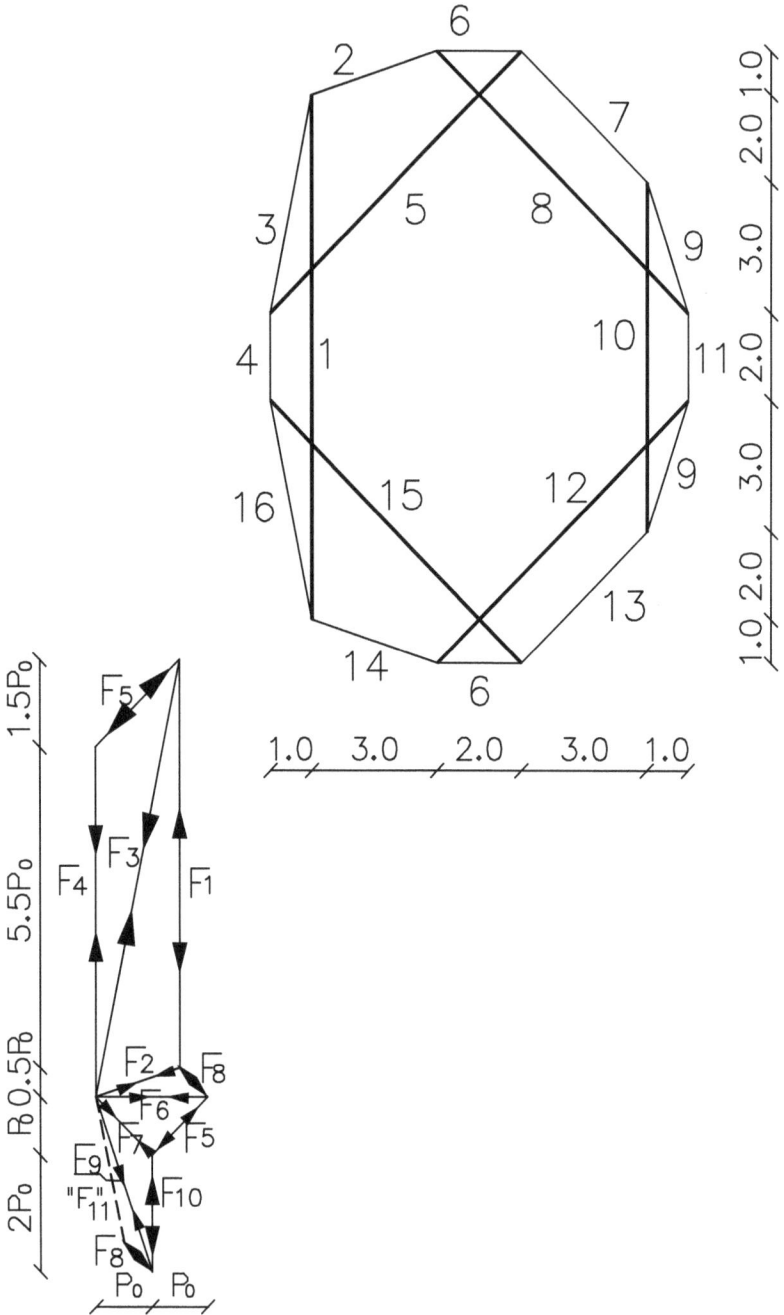

Figure 5.10 Symmetrical tensegrity ring with six bars.

Part 2

Three-dimensional spatial tensegrity structures

Chapter 6

Tensegrity slabs

6.1 SPATIAL TENSEGRITY SANDWICHES

A spatial tensegrity sandwich comprises two layers of cable nets with bars between them. The tensegrity nets shown in Chapter 2 are used to form the spatial tensegrity sandwich. These tensegrity nets form the top and bottom layers of the sandwich. The bars connecting these two layers of cables are of H shape and elaborate H shape is shown in Figure 6.1.

These bars are arranged as shown in the various tensegrity nets presented in Chapter 2. An example of a plan of a spatial tensegrity sandwich formed from the tensegrity net shown in Figure 2.10 is shown in Figure 6.2.

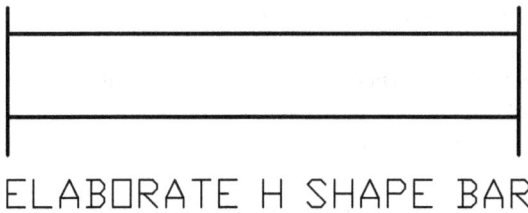

Figure 6.1 Shape of the bars in the spatial tensegrity sandwich.

DOI: 10.1201/9781003370093-8

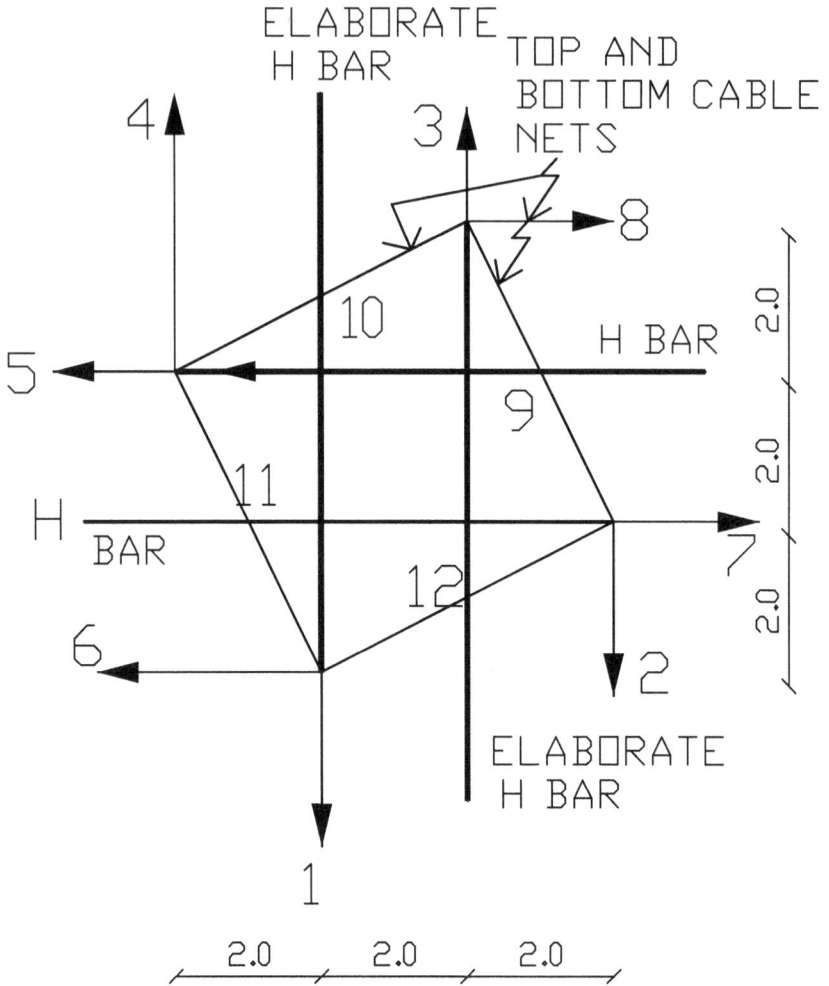

Figure 6.2 Plan of a typical section of the tensegrity spatial sandwich.

A three-dimensional view of a section of this spatial tensegrity sandwich is shown in Figure 6.3.

By using this method, other tensegrity nets presented in Chapter 2 can be used to form other types of spatial tensegrity sandwiches.

Prestressing the spatial tensegrity sandwich can follow the methods presented in Section 2.2.

Because this spatial tensegrity sandwich can be appropriately prestressed, it is a feasible tensegrity structure and it can be constructed.

ELABORATE H BAR

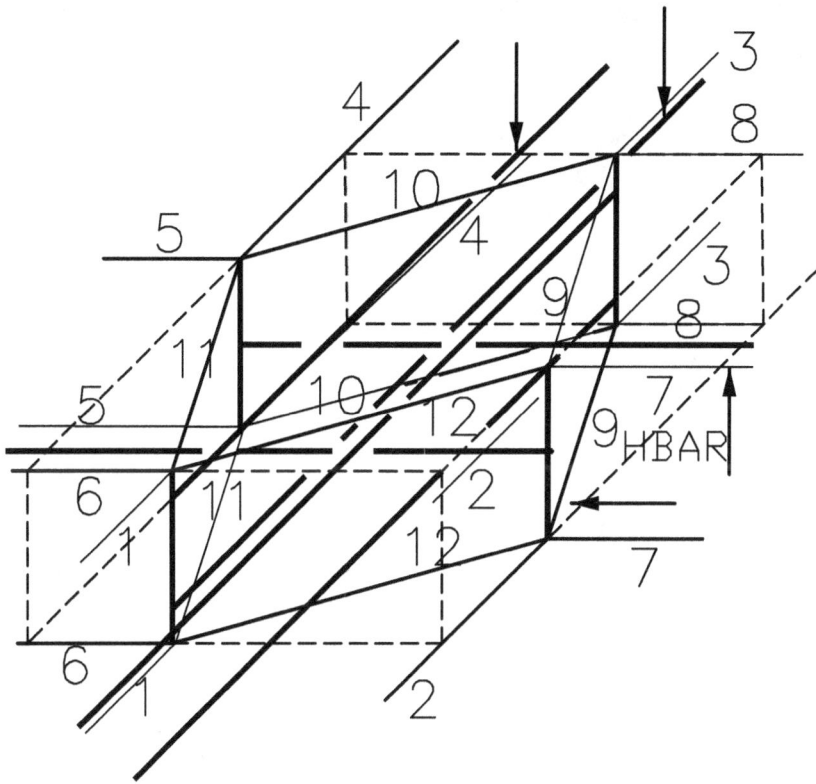

Figure 6.3 View of a typical section of the tensegrity spatial sandwich.

6.2 ELABORATE TENSEGRITY SLABS

The elaborate tensegrity slab is an intricate structure. It consists of individual cells connected by cables and bars to each other. The cells are composed of cables forming tilted cubes. Each cell is composed of two squares, one at the top of the cell and one at the bottom. The square at the top of each cell is tilted with regard to the bottom one. The nodes of the squares at the top and bottom are connected with cables. The exact nature of these cables is discussed later. The arrangement of the cells squares is shown in Figure 6.4. The nodes of the cables at the top square are marked by 1, 2, 3 and 4. The nodes of the cables at the bottom square are marked by 5, 6, 7 and 8.

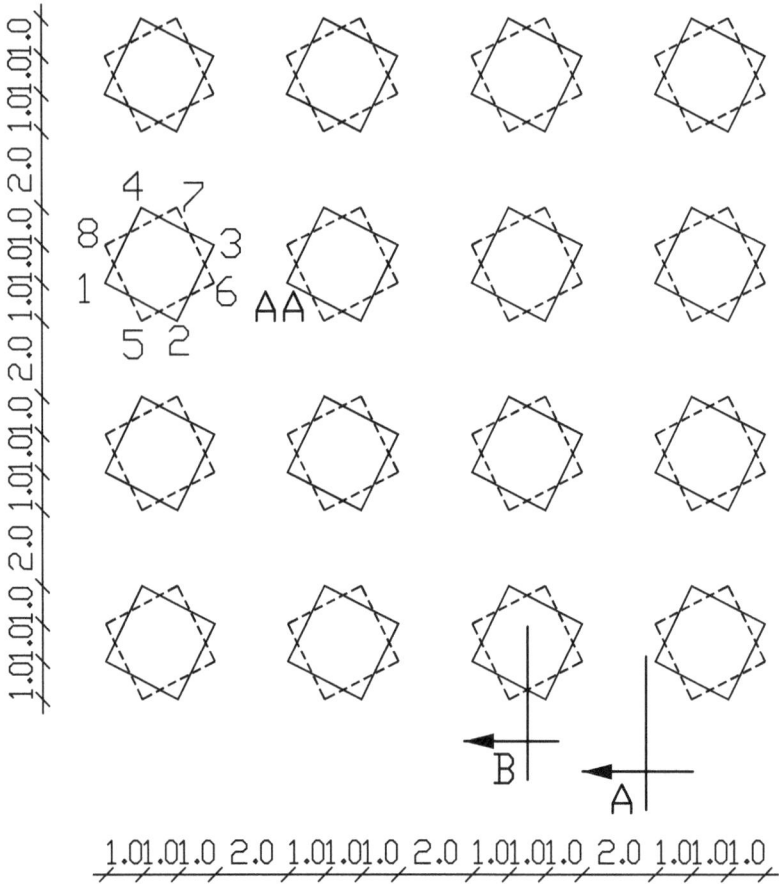

Figure 6.4 Arrangement of the squares of the tensegrity elaborate slab.

The cells are connected to each other by bars and by two families of cables. The bars marked by element No. 1 are shown in Figure 6.5. There are two families of cables. The first family of cables marked by element No. 2 is shown in Figure 6.6. The second family of cables marked by element No. 3 is shown in Figure 6.7. Cross sections A:A and B:B are shown in Figure 6.8. The slab thickness is up to the designer and it is assumed to be d units of the units used in the plan of the elaborate tensegrity slab.

The slab is prestressed by inducing force into the bars of element No. 1. It is assumed that the vertical component of this prestressing force, F^1_z, is equal to P_0.

Equilibrium of the vertical forces at cross section C:C shown in Figures 6.5-6.8 takes the form shown in Figure 6.9.

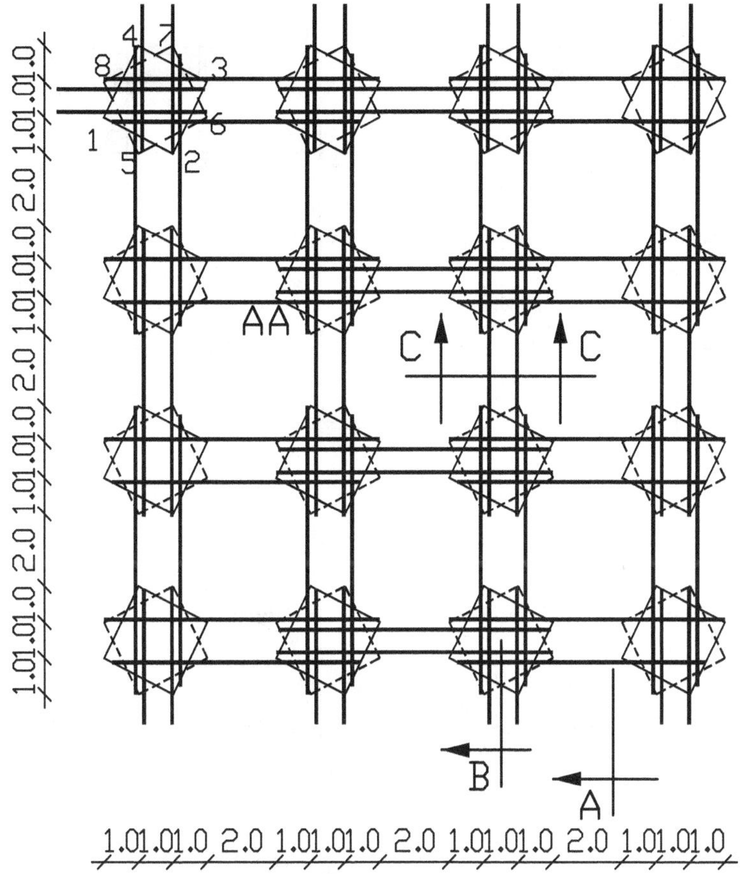

Figure 6.5 Arrangement of the bars (element No. 1).

Equilibrium of moments implies:

$$6.1 \quad P_0 = 9F_z^2 + 2F_z^3$$

Here, F_z^2 and F_z^3 are the vertical components of the force in the cables of element No. 2 and the elements of cables No. 3, respectively.

Since

$$6.2 \quad \tan\alpha = d/8; \quad \tan\beta = d/14; \quad \tan\mu = d/13$$

horizontal equilibrium of the forces acting at cross section C:C implies:

$$6.3 \quad 8P_0 = 42F_z^2 + 26F_z^3$$

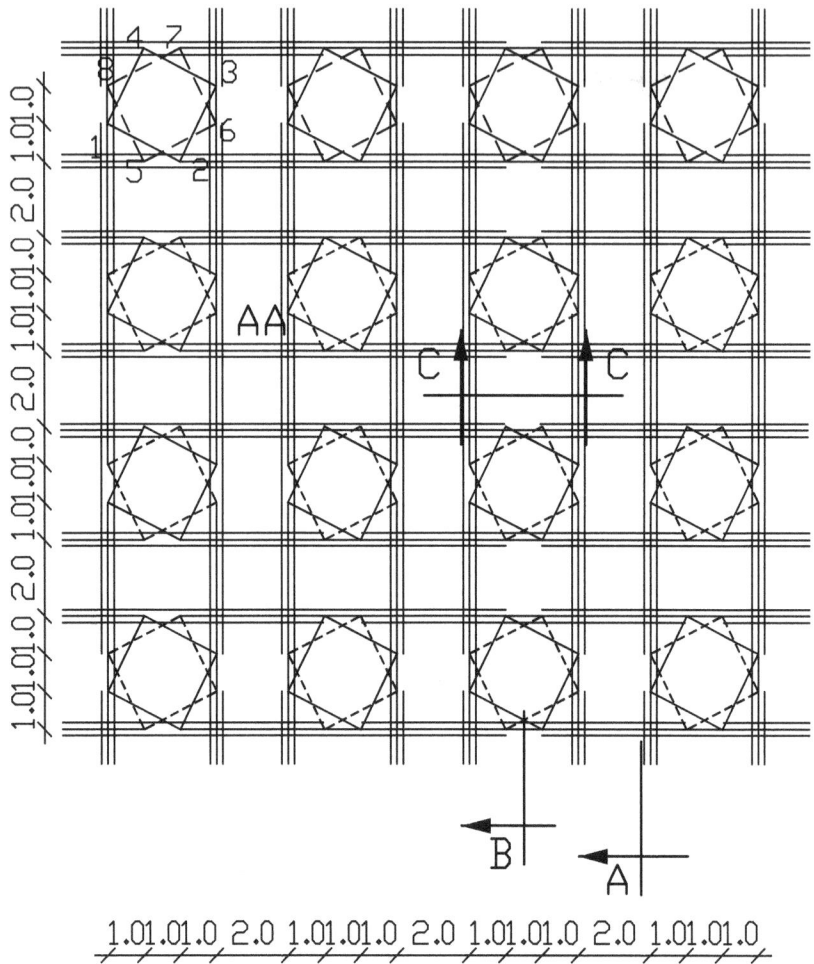

Figure 6.6 Arrangement of the first family of cables (element No. 2).

By using Equations 6.1 and 6.3, the forces at the cables of elements No. 2 and the cables of elements No 3 are as follows:

$$6.4 \quad \begin{aligned} F_z^2 &= P_0 / 15 \\ F_z^3 &= P_0 / 5 \end{aligned}$$

Bars of element No. 1 and cables of elements No. 2 and cables of element No. 3 apply the forces F_x, F_y and F_z shown in Figure 6.10 to a typical node AA, shown in Figures 6.4–6.7.

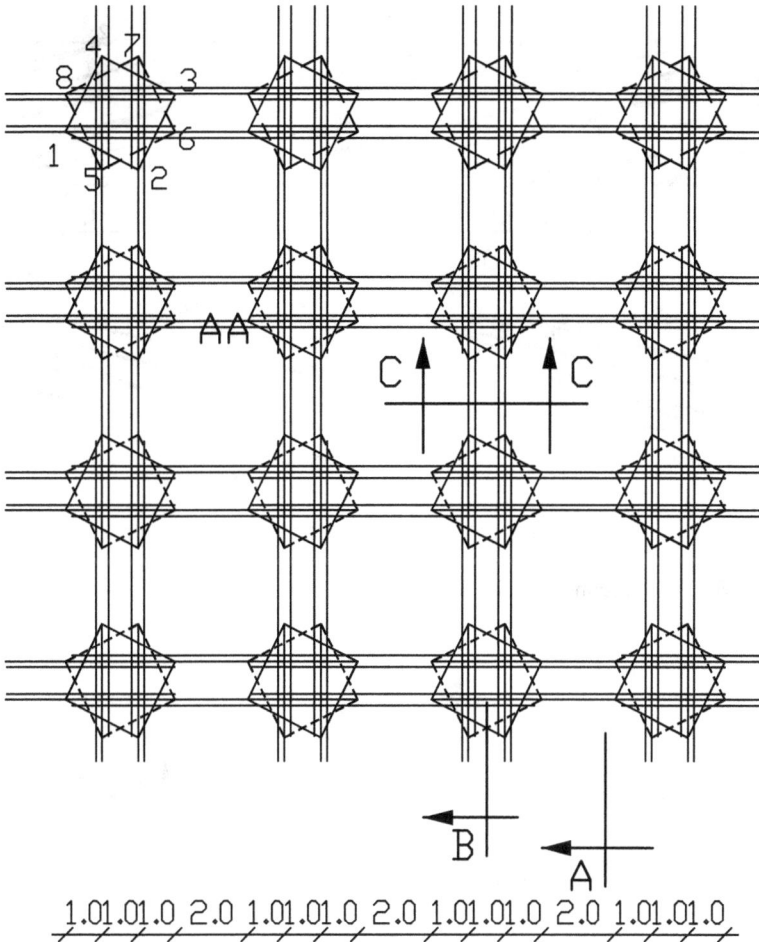

Figure 6.7 Arrangement of the second family of cables (element No. 3).

It is easy to realize that

$$F_z = F_z^1 - F_z^2 - F_z^3 = 11P_0/15$$

$$6.5 \quad F_x = F_x^1 - F_x^3 = 27P_0/(5d)$$

$$F_y = F_y^2 = 14P_0/(15d)$$

The footprint of the resultant of the forces applied, by bars of element No. 1 and cables of element No. 2 and cables of element No. 3, at node AA at the

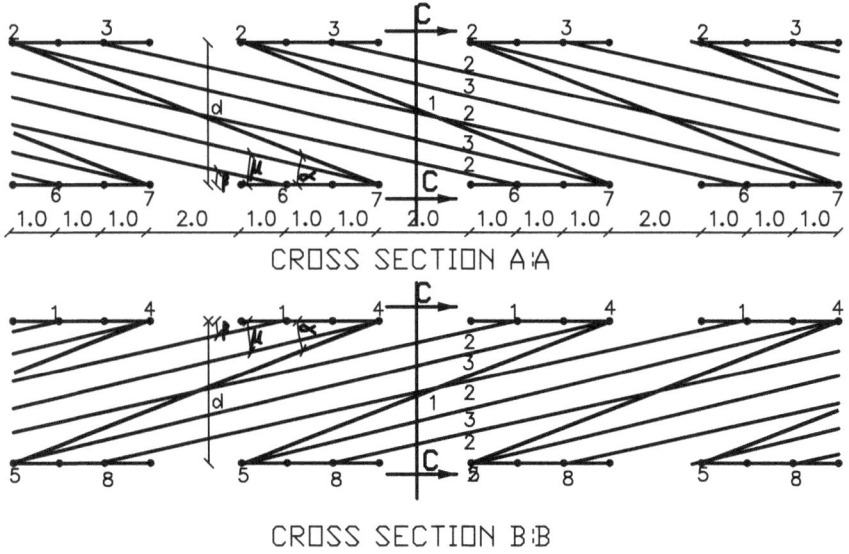

CROSS SECTION A:A

CROSS SECTION B:B

Figure 6.8 Cross sections A:A and B:B.

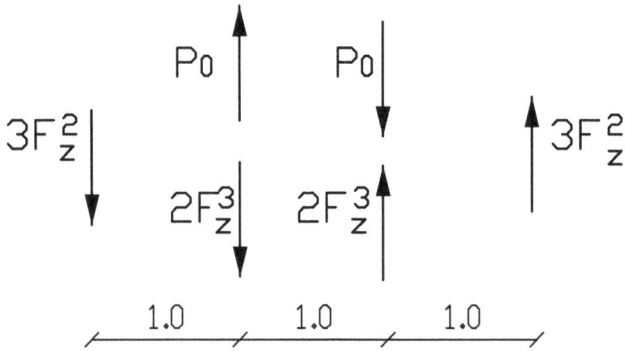

Figure 6.9 The forces at cross section C:C.

bottom level of the elaborate tensegrity slab is at point B at a distance of L_x^B and L_y^B shown in Figure 6.10.

$$6.6 \quad \begin{aligned} L_x^{\ B} &= dF_x \, / \, F_z = 71/11 = 6.45 \\ L_y^{\ B} &= dF_y \, / \, F_z = 14/11 = 1.27 \end{aligned}$$

Because of the symmetry of the squares of the elaborate tensegrity slab, the forces at the four top cables that form the square at the top should be

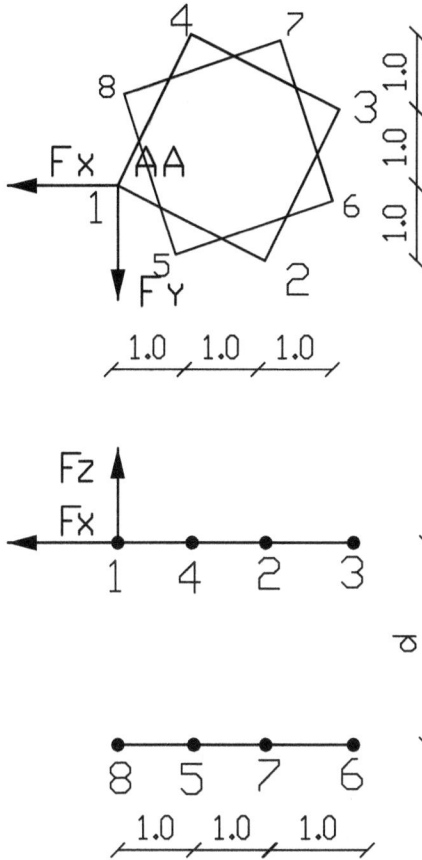

Figure 6.10 Forces applied by bar 1 and cables 2 and 3 to a typical node AA of the cell.

the same. Thus, the resultant of these forces at node AA is horizontal and is acting along the line passing through points 1 and 3 as shown by the dotted line in Figure 6.11.

To ensure equilibrium of forces at the nodes of the top square, each node should be connected by cables to at least two joints in the opposite square at the bottom. In the design of these cables, care should be taken to ensure that prestressing induces tension to them. For example, if node 1 is connected to nodes 5 and 8 by cables, the footprint of the forces acting at point B is within the area at the bottom level bordered by these cables as shown in Figure 6.11 and thus all cables are in tension. Following nodes 1 2 should be connected to nodes 5 and 6 and so on as it is shown in Figure 6.12.

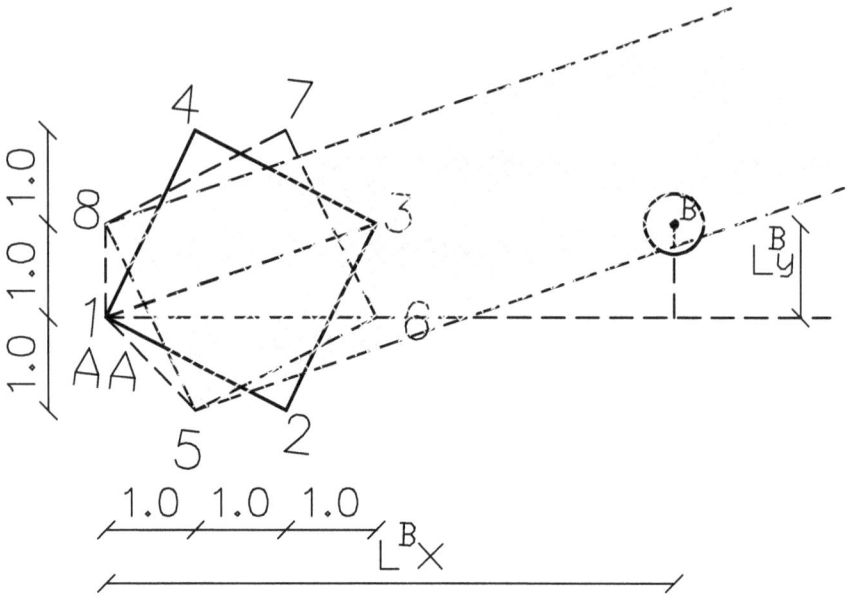

Figure 6.11 Footprint of the resultant of the forces at node AA.

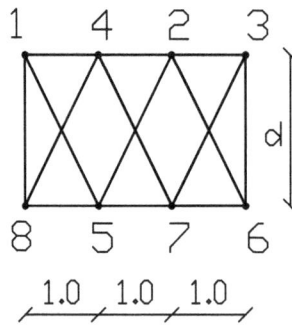

Figure 6.12 Cables between the top and bottom nodes.

Obviously, there are other suitable options available to the designer in the design of these cables.

Because this elaborate tensegrity slab is appropriately prestressable, it is a feasible tensegrity structure.

It is proposed to construct this elaborate tensegrity slab in the following way. At first, prestress two cable nets of squares one at the top level and one at the bottom level as shown in Figure 6.13.

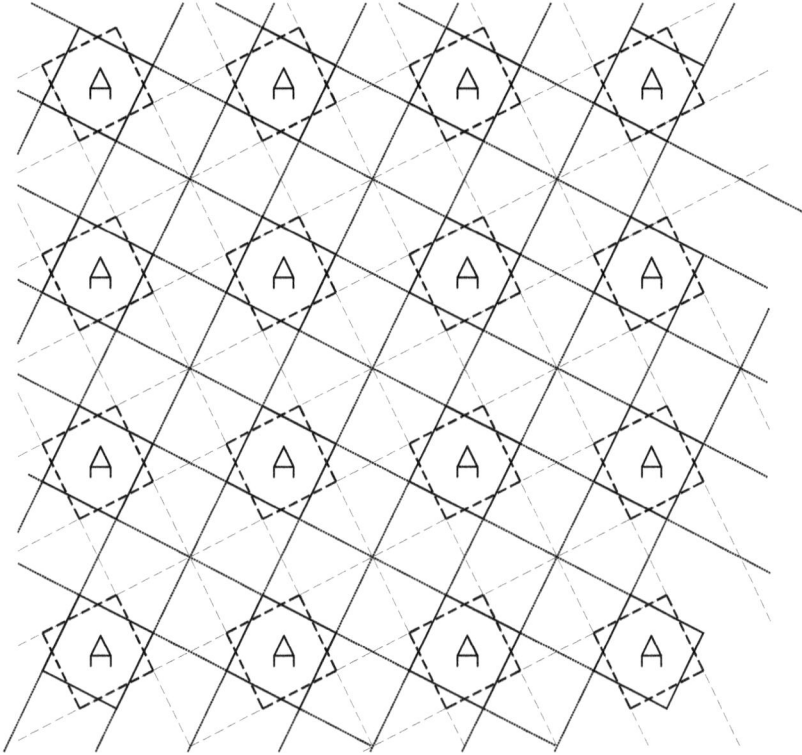

Figure 6.13 Top and bottom cable nets for the construction of the elaborate tensegrity slab.

The cable nets are designed according to the squares of the elaborate tensegrity slab shown in Figures 6.4 and are marked by A. Then the bars shown in Figure 6.5 and the two families of cables shown in Figures 6.6 and 6.7 are fitted in. Also, the cables shown in Figure 6.12 should be installed too. At the final stage, the redundant cables at the top and bottom levels, which are not required for the configuration of the elaborate tensegrity slab, are to be removed.

Chapter 7

Spatial tensegrity rings

The plane tensegrity ring shown in Figure 5.1 can be used to form a spatial tensegrity ring as shown in Figure 7.1.

Forces in bars 1, 7, 4 and 8 and in cables 3, 6, 12 and 10 are shown in the force diagram in Figure 5.2. Forces V are required to maintain the spatial tensegrity ring in equilibrium. Each V is function of force F_2 in cables 2, 5, 9 and 11 shown in Figure 5.2 and takes the following form:

$$7.1 \quad V = F_2 h / l$$

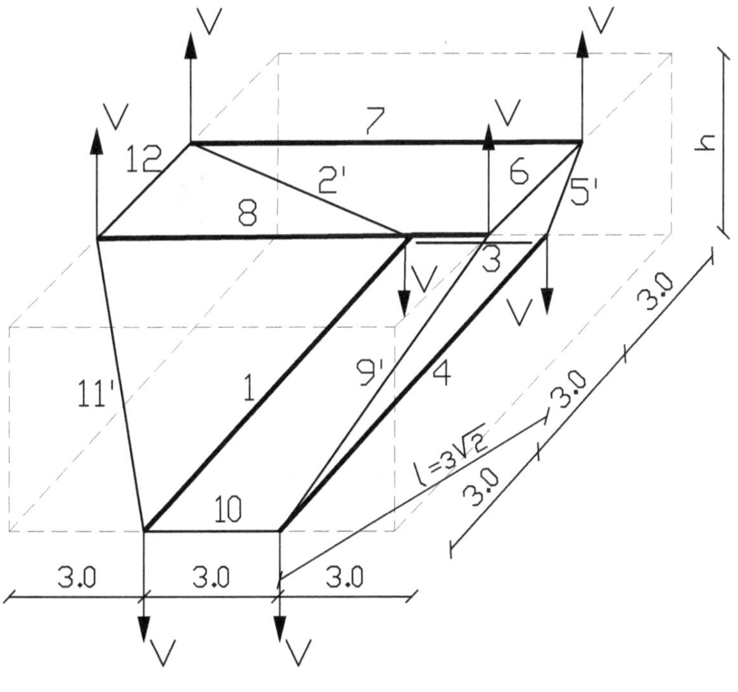

Figure 7.1 View of a spatial tensegrity ring based on the tensegrity ring shown in Figure 5.1.

DOI: 10.1201/9781003370093-9

Here, $l = 3\sqrt{2}$.

This force implies that the actual force F'_2 induced by the prestressing of the spatial tensegrity ring to cables 2', 5', 9' and 11' shown in Figure 7.1 is as follows:

$$7.2 \quad F'_2 = F_2 (l^2 + h^2)/l$$

In the case where the spatial tensegrity ring is part of a tensegrity structure, force V indicates the forces applied by the spatial tensegrity ring to the rest of the structure.

It is possible to add spatial tensegrity rings one on top of the other to form a tensegrity tower as shown in Figure 7.2.

In the case of the tensegrity tower forces in bars 1, 7, 4 and 8 and in cables 3, 6, 12 and 10 are double the forces given in the force diagram shown in Figure 5.2. Forces in cables 2', 5', 9' and 11' are given in Equation 7.2.

Force V applied to the supporting frame is given in Equation 7.1.

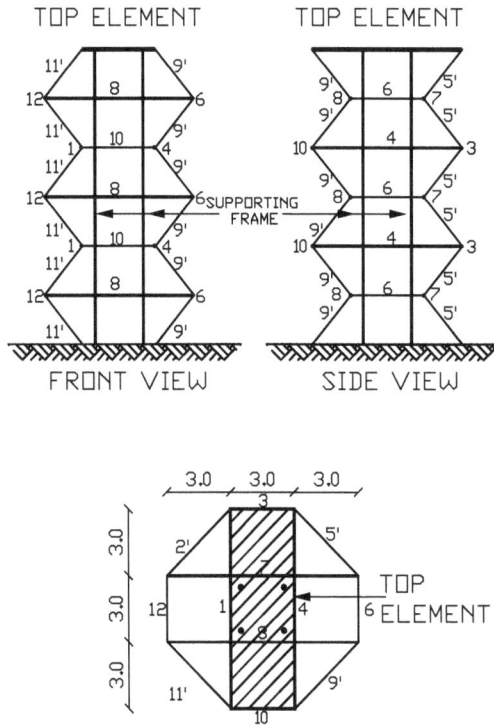

Figure 7.2 Tensegrity tower.

Chapter 8

Tensegrity vaults

8.1 LEVEL TENSEGRITY VAULTS

The tensegrity arch shown in Figure 4.8 can be used to construct a level tensegrity vault. In the proposed level tensegrity vault, the appropriate cables are parallel to the bars at equal spacing as shown in Figure 8.1.

A plan of the level tensegrity vault is shown in Figure 8.2 and a side view in Figure 8.3.

Forces in bars 1, 4 and 7 and in cables 8, 5 and 2 are shown in the force diagram shown in Figure 8.1 and are equal to P_0. The forces in cables 3′ and 6′ are function of the forces F_3 and F_6 given in the force diagram shown in Figure 8.1.

$$8.1 \quad F_3 = F_6 = P_0 \sqrt{2 + \sqrt{2}}$$

Since two cables 3′ and 6′ replace cables 6 and 3, $F_3′$ and $F_6′$ take the following form:

$$8.2 \quad F_3' = F_6' = P_0 \sqrt{b^2 + 2(2 + \sqrt{2})a^2 / (a2\sqrt{2})}$$

Also, $\tan \alpha$ takes the form given by Equation 8.3:

$$8.3 \quad \tan \alpha = b / a$$

By using Equation 8.3, Equation 8.2 takes the following form:

$$8.4 \quad F_3' = F_6' = P_0 \sqrt{2(\tan \alpha)^2 + 4(2 + \sqrt{2})} / 4$$

Equation 8.4 indicates that the magnitudes of $F_3′$ and $F_6′$ are function of α.

DOI: 10.1201/9781003370093-10

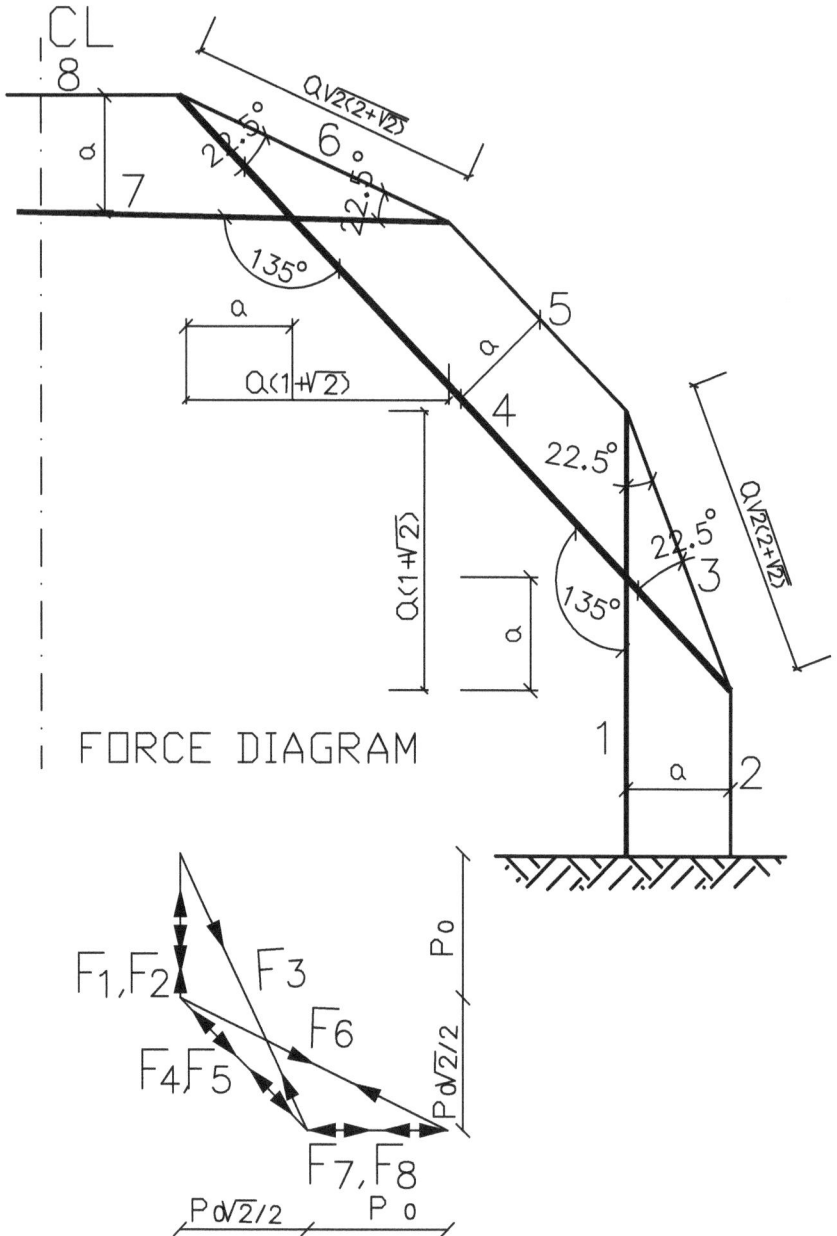

Figure 8.1 Side view of the level tensegrity vault.

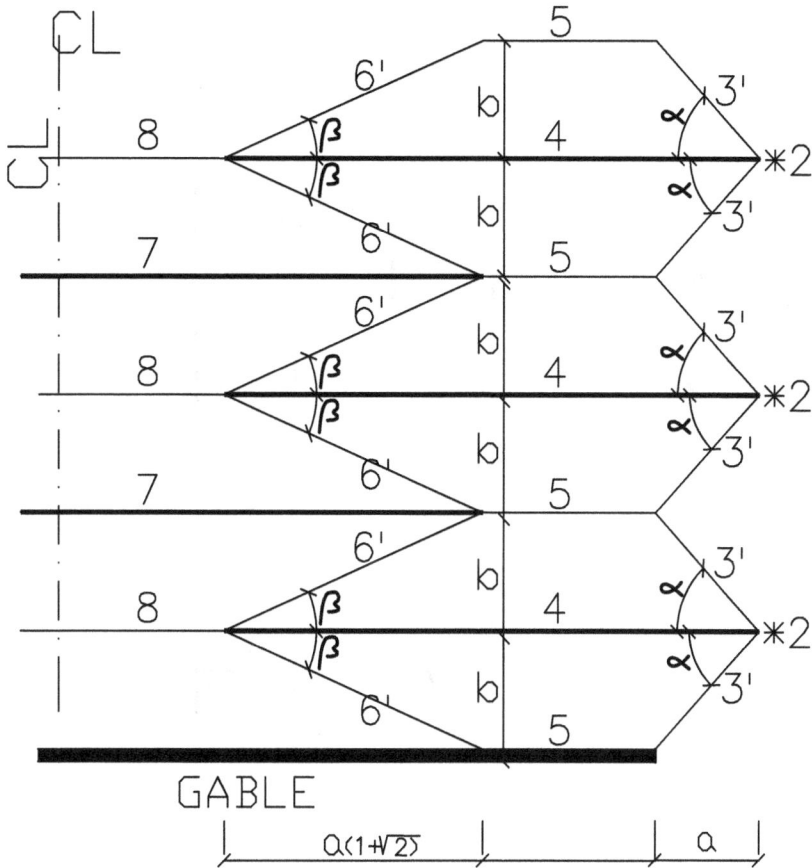

Figure 8.2 Plan of the levelled tensegrity vault.

The side gables can have a large foundation to negotiate with the moment applied to them by the tensegrity vault, or, if there is enough space outside the structure perimeter, they can be supported by cables as shown in Figure 8.3.

The levelled tensegrity vault demonstrates an important feature of tensegrity structures. In common trusses, the number of members should be, at least, equal to the number of equilibrium equations required to determine the forces in the truss members. In tensegrity structures, the number of elements can be less than this number.

Various methods can be used to erect the level tensegrity vault. A proposed method is shown in Figure 8.4.

The method is based on using additional structural elements to construct free independent units of the level tensegrity vault. The units are marked

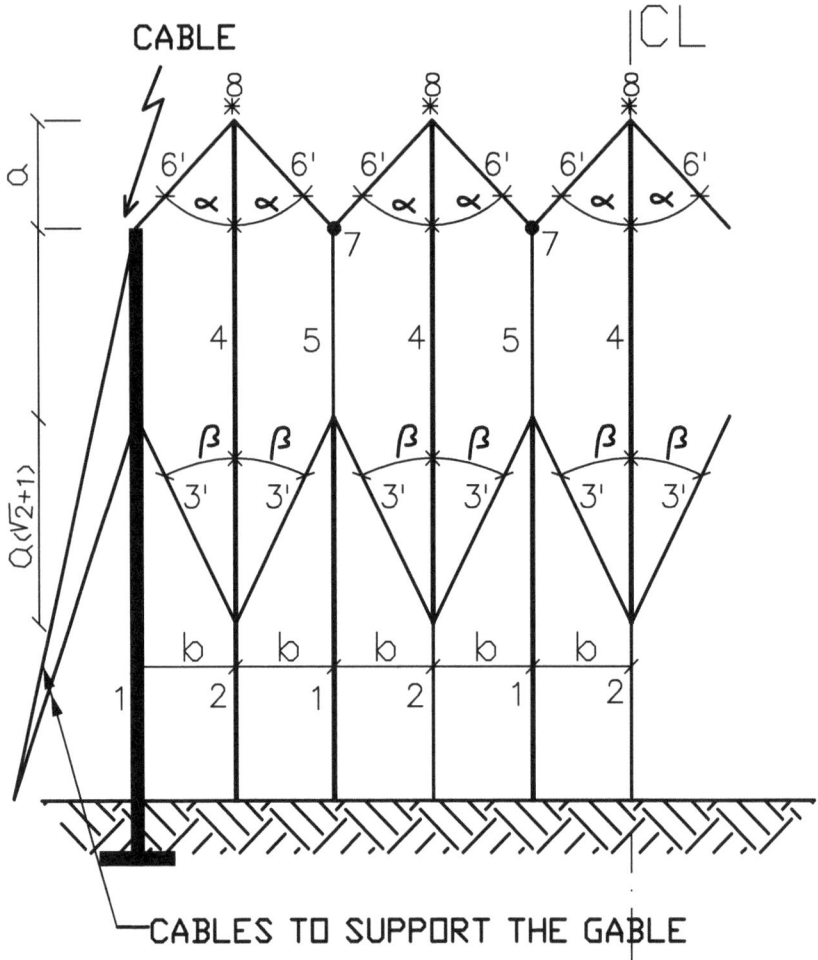

Figure 8.3 Front view of the level tensegrity vault.

by A, B and C and shown in Figure 8.4. The additional structural elements are shown by dotted lines. The additional structural elements are designed so that they can be removed easily when necessary. Each unit is prestressed to a minimum level to stabilize it only. Scaffolding are used to arrange the units in place to form the first layer shown in Figure 8.4. After the first layer is placed, the second layer shown in Figure 8.4 can be placed. Care should be taken to join together the identical nodes at the different units to each other. Some cables and bars of the level tensegrity vault will be composed of two members of two adjacent units. After the tensegrity vault was constructed, it should be prestressed to the minimum level required to remove

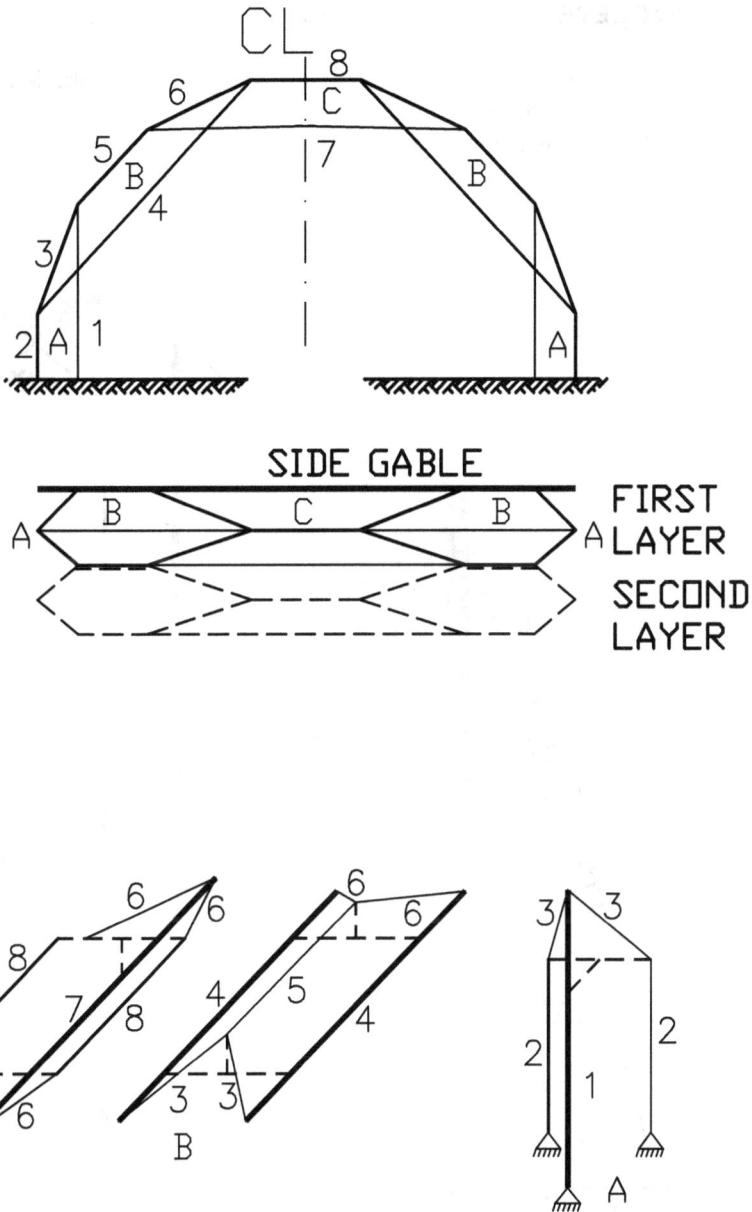

Figure 8.4 Proposed method to construct the level tensegrity vault.

the additional structural elements from all the units. After removing the additional structural elements, the level tensegrity vault can be prestressed to the designed level.

8.2 NOT LEVEL TENSEGRITY VAULTS

Following the method used to design the level tensegrity vault, described in Section 8.1, it is possible to design a not level tensegrity vault, shown in Figures 8.5–8.7. In this case, the designer is free to fix the locations of nodes a, c

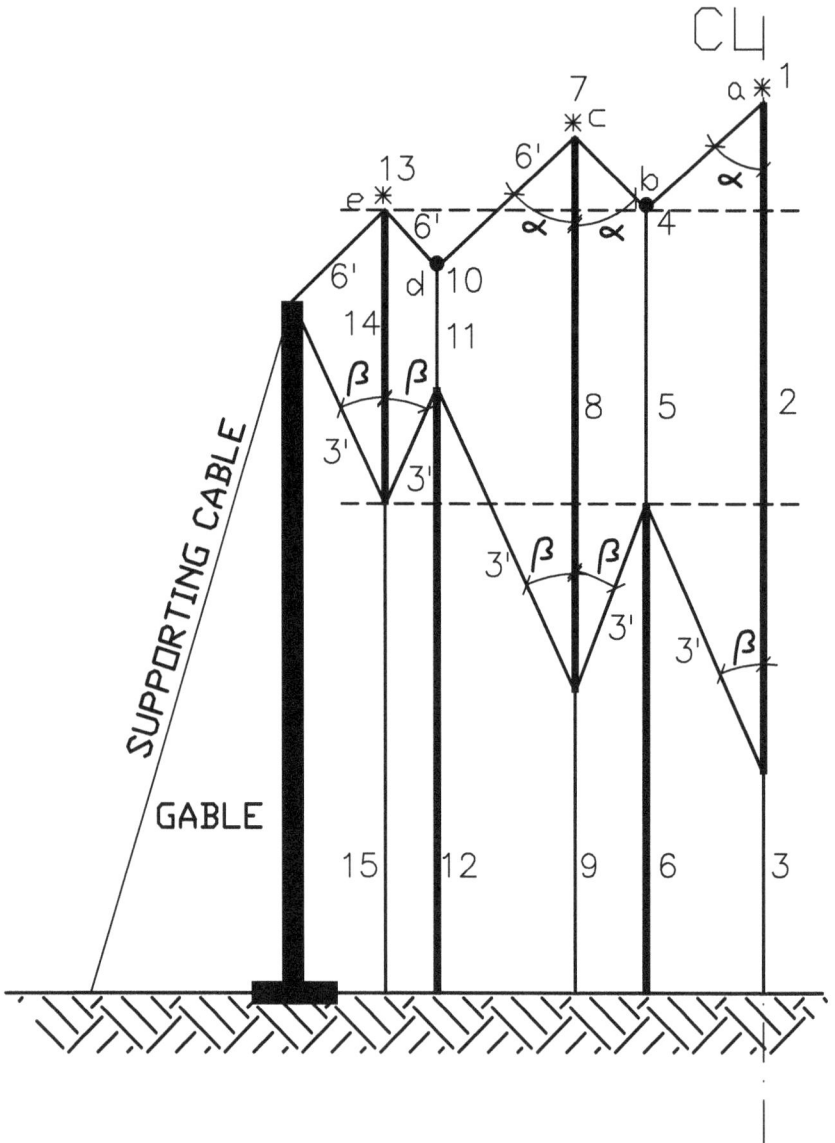

Figure 8.5 Front view of the not level tensegrity vault.

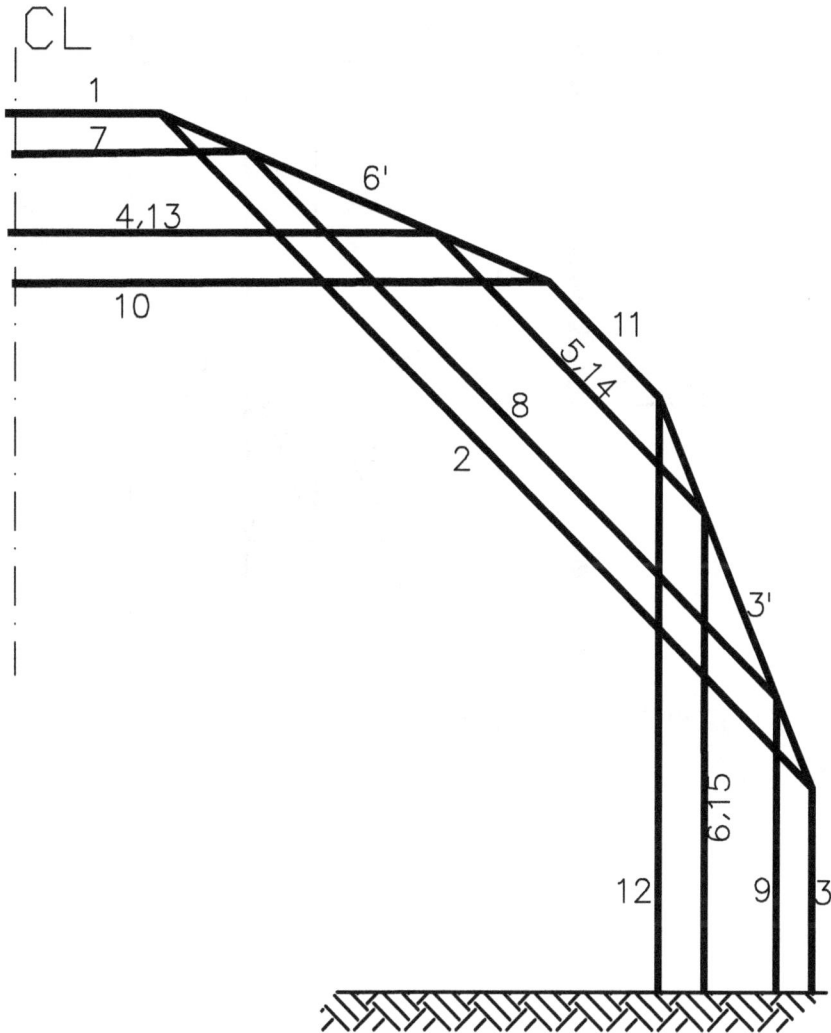

Figure 8.6 Side view of the not level tensegrity vault.

and e shown in Figure 8.5. By keeping the angel α as shown in Figure 8.5, the locations of nodes b and d can be determined. The rest of the nodes are determined by keeping angel β shown in Figure 8.5. Following this method, the side view of the not level tensegrity vault takes the form shown in Figure 8.6 and the plan of the not level tensegrity vault takes the form shown in Figure 8.7.

Forces induced by prestressing to cables 1, 3, 7, 9, 13 and 15 is P_0 tension and to bars 2, 4, 6, 8, 10, 12 and 14 is P_0 compression as shown in Figure 8.1. The tension in cables 3′and 6′ is given by Equation 8.4.

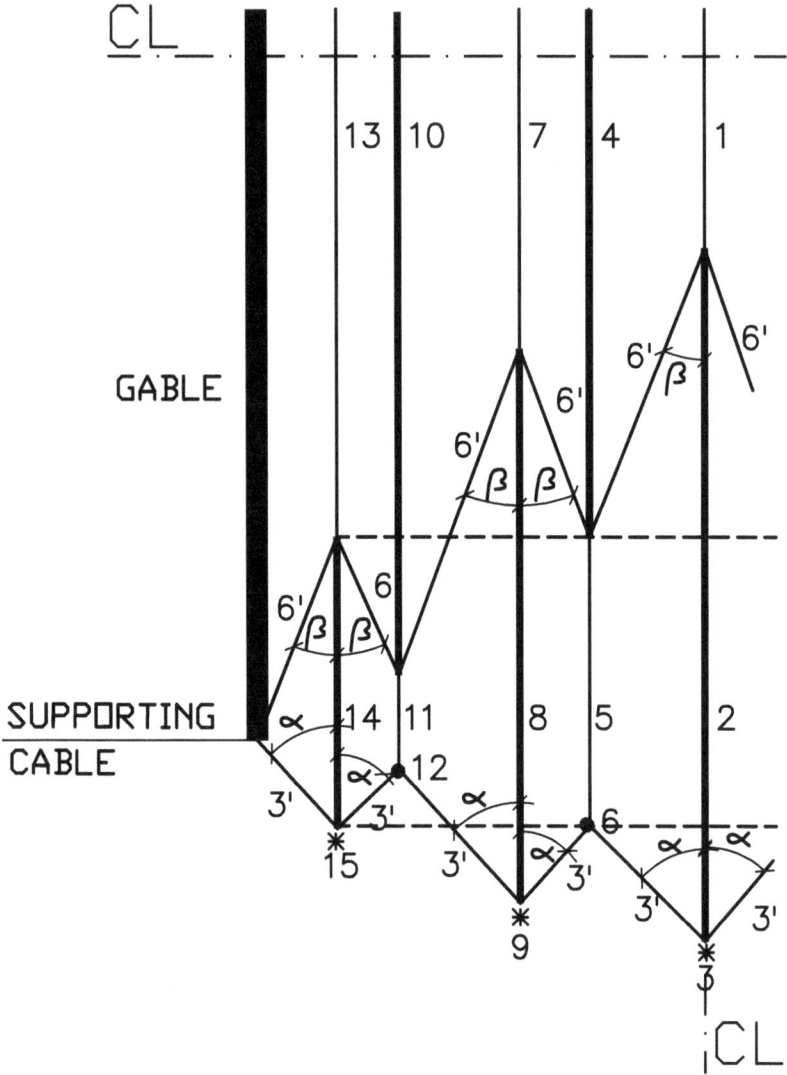

Figure 8.7 Plane of the not level tensegrity vault.

Also, in this case, erection can follow the method described in the case of the level tensegrity vault.

Chapter 9

Tensegrity cylindrical shells

9.1 VERTICAL TENSEGRITY CYLINDRICAL SHELLS

The tensegrity arch shown in Figure 8.1 and the tensegrity net shown in Figure 2.10 are used to form the vertical tensegrity cylindrical shell. Plan of the vertical tensegrity cylindrical shell is shown in Figure 9.1 and a side view in Figure 9.2.

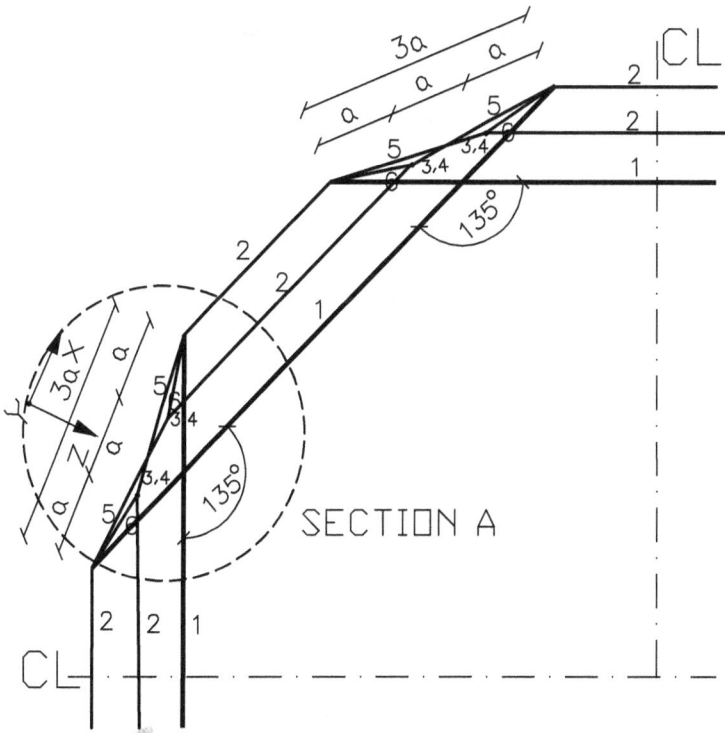

Figure 9.1 Plan of the vertical tensegrity cylindrical shell.

DOI: 10.1201/9781003370093-11

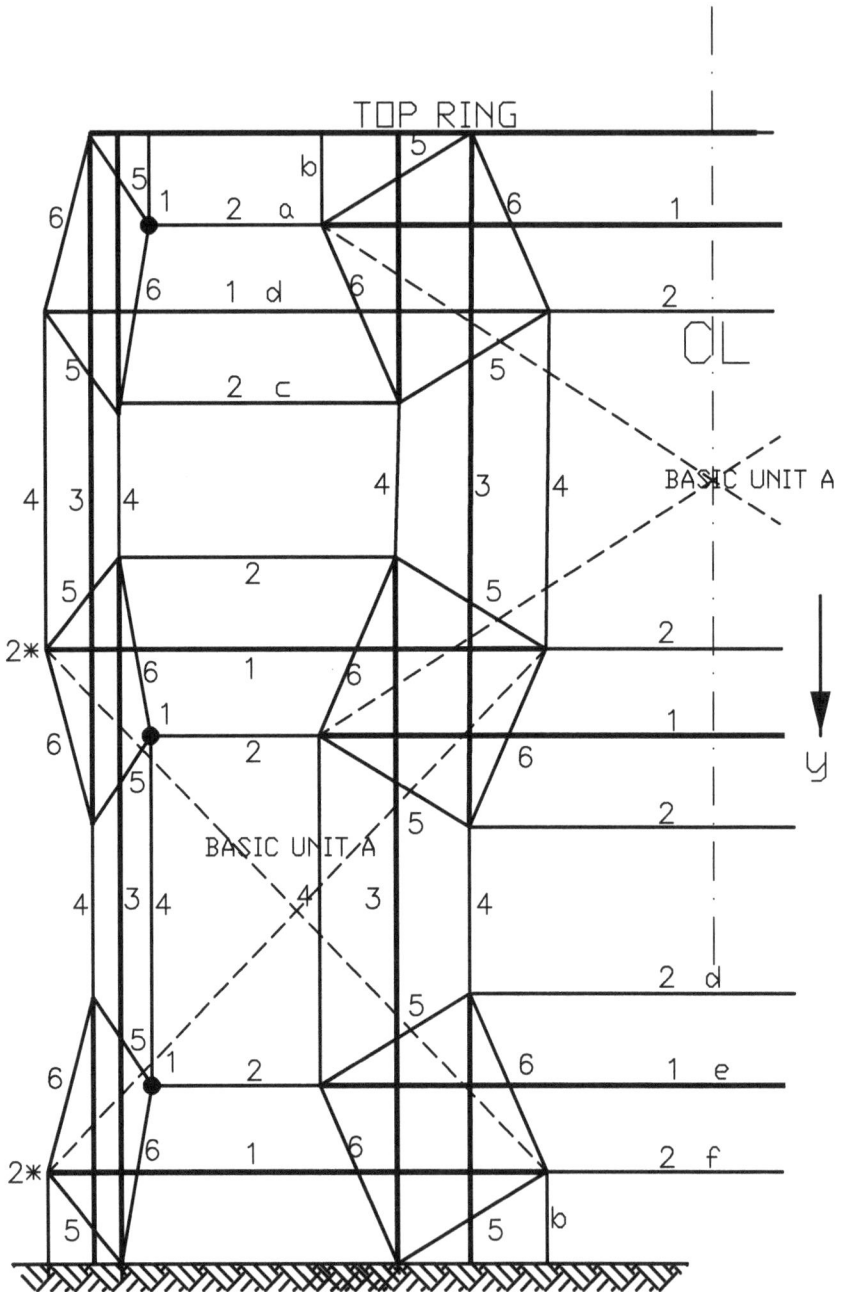

Figure 9.2 Side view of the vertical tensegrity cylindrical shell.

Section A of the vertical tensegrity shell shown in Figure 9.1 is shown in Figure 9.3.

Prestressing applies force P_0 to the component of F^1_x and F^3_y (in absolute values):

$$9.1 \quad F_x^1 = F_y^3 = P_o$$

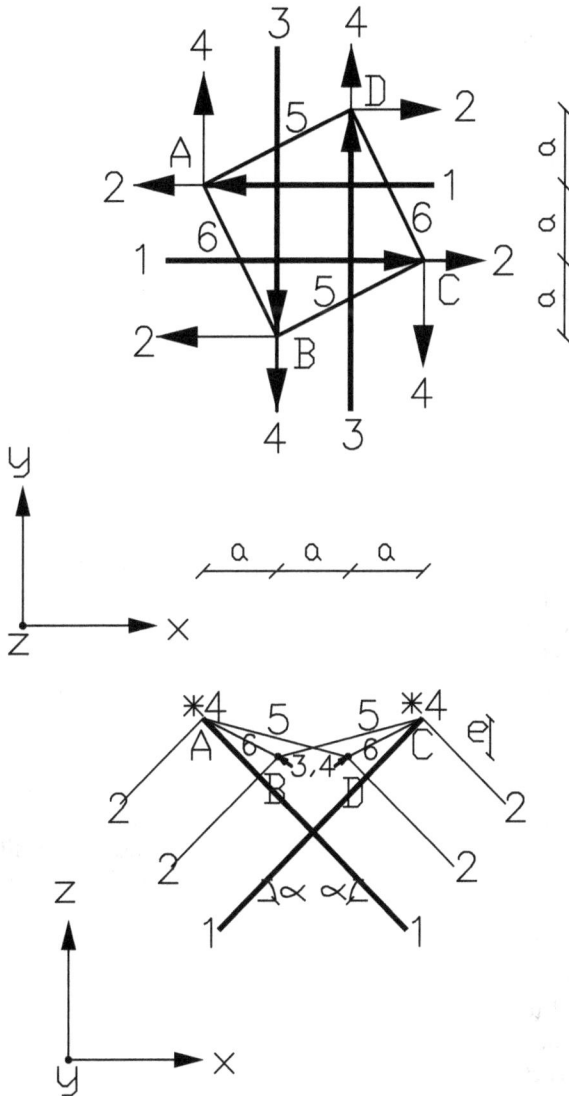

Figure 9.3 Prestressing forces at section A.

By using Figure 2.11 equilibrium of section A implies:

$$9.2 \quad F^2_{\,x} = F^4_{\,y} = P_0 / 2$$

$$9.3 \quad F^5_{\,y} = F^6_{\,x} = P_0 / 2$$

$$9.4 \quad F^6_{\,y} = F^5_{\,x} = P_0$$

Equilibrium at nodes A and C in the z-direction implies:

$$9.5 \quad F^1_{\,z} - F^2_{\,z} = F^6_{\,z} + F^5_{\,z} = P_0 \tan\alpha / 2$$

Vertical equilibrium at nodes B and D implies:

$$9.6 \quad F^2_{\,z} = F^6_{\,z} + F^5_{\,z} = P_0 \tan\alpha / 2$$

Since $F^6_{\,z} = P_0 e / (2a)$ and $F^5_{\,z} = P_0 e / (2a)$

$$9.7 \quad F^6_{\,z} + F^5_{\,z} = P_0 e / a$$

By using Equations 9.6 and 9.7, e takes the following form:

$$9.8 \quad e = a \tan \alpha / 2$$

The fact that the vertical tensegrity cylindrical shell can be appropriately pre-stressed in the proposed configuration indicates that it is a feasible tensegrity and can be constructed in this configuration.

The method of constructing the vertical tensegrity shell is up to the designer. It is proposed to construct the vertical tensegrity cylindrical shell by using a basic prestressed tensegrity unit A shown in Figure 9.4.

Each unit constructs with additional structural elements shown by the dotted lines in Figure 9.4 and prestressed to a minimum level to stabilize it. The units are placed in a chessboard pattern as shown in Figure 9.2. Care should be taken to join together the identical nodes at the different units to each other. So some cables and bars of the vertical tensegrity cylindrical shell are composed of two members of two adjacent units. After the vertical cylindrical shell is assembled, the redundant elements along the boundaries marked by a, b, c, d, e and f shown in Figure 9.2 are added. They should be prestressed to a minimum level to ensure that all cables are tight. Then prestressing the vertical cylindrical shell to the level required to remove the additional structural elements from all the units. After removing the additional structural elements, the vertical cylindrical shell can be prestressed to the designed level.

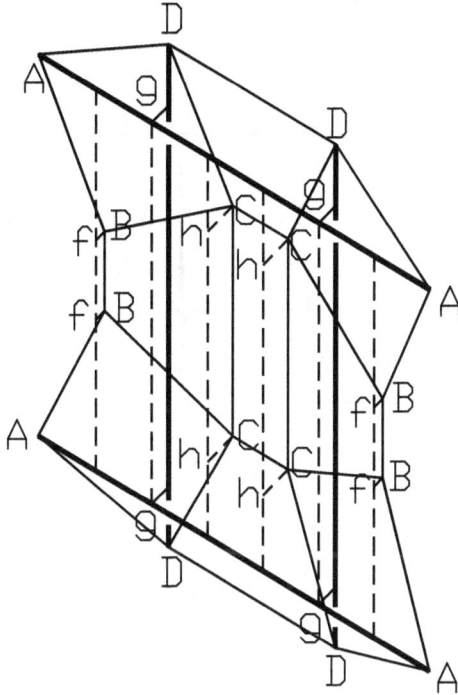

Figure 9.4 Basic unit A.

9.2 HORIZONTAL TENSEGRITY CYLINDRICAL SHELLS

The tensegrity cylindrical shell can be constructed horizontally as shown in Figures 9.5–9.7. In this case, the forces induced by prestressing are the same as those analysed in the case of a vertical tensegrity cylindrical shell presented in Section 9.1.

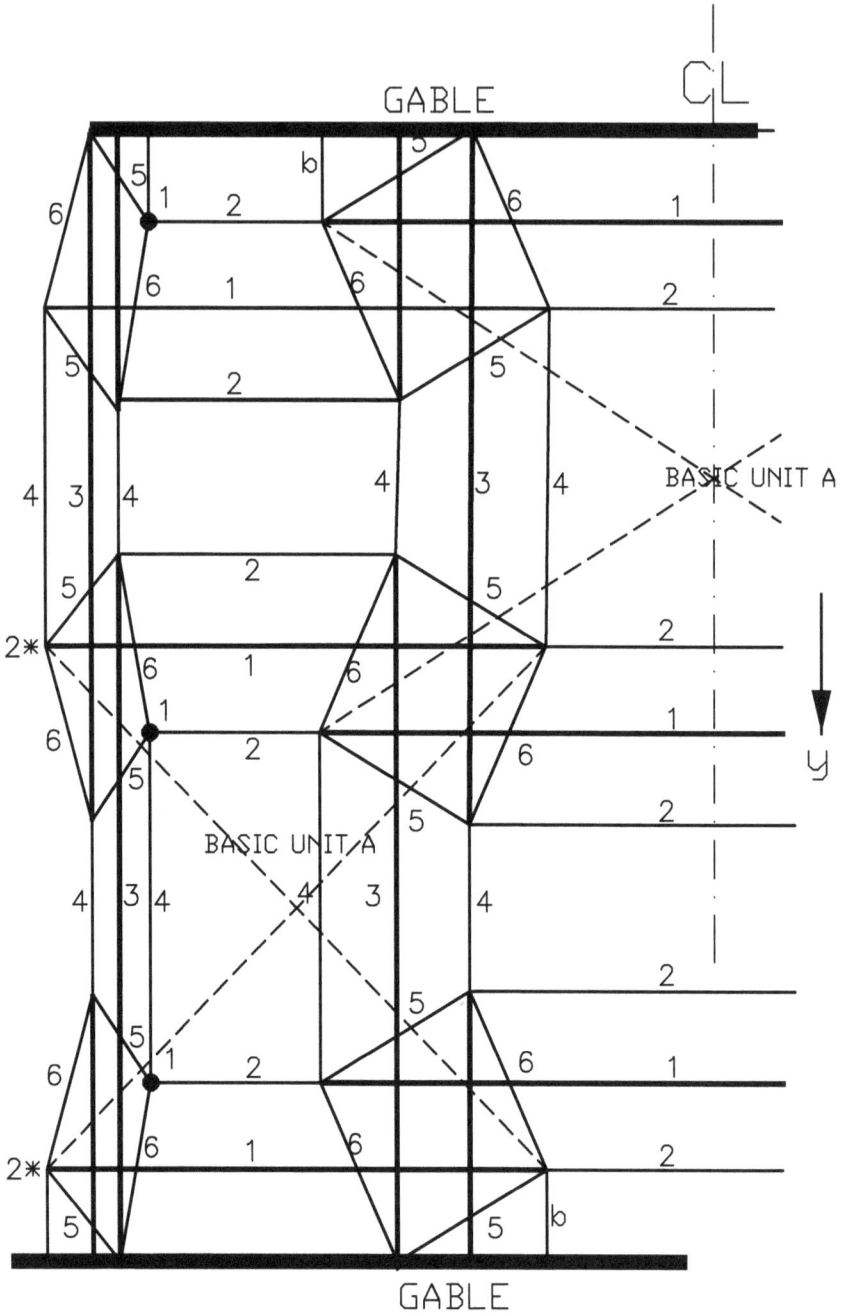

Figure 9.5 Plan of a horizontal tensegrity cylindrical shell.

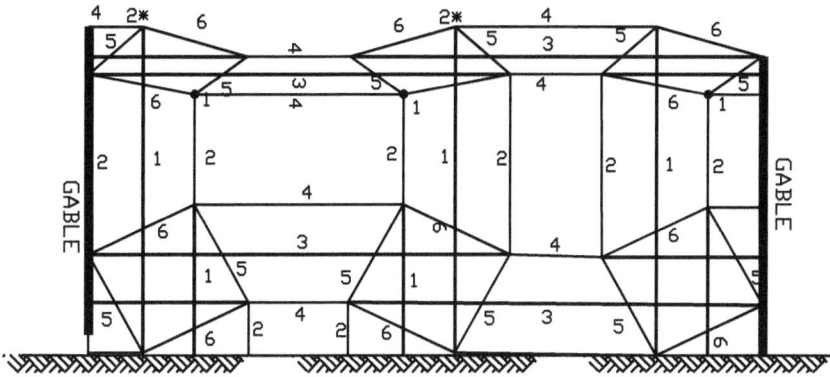

Figure 9.6 Side view of a horizontal tensegrity cylindrical shell.

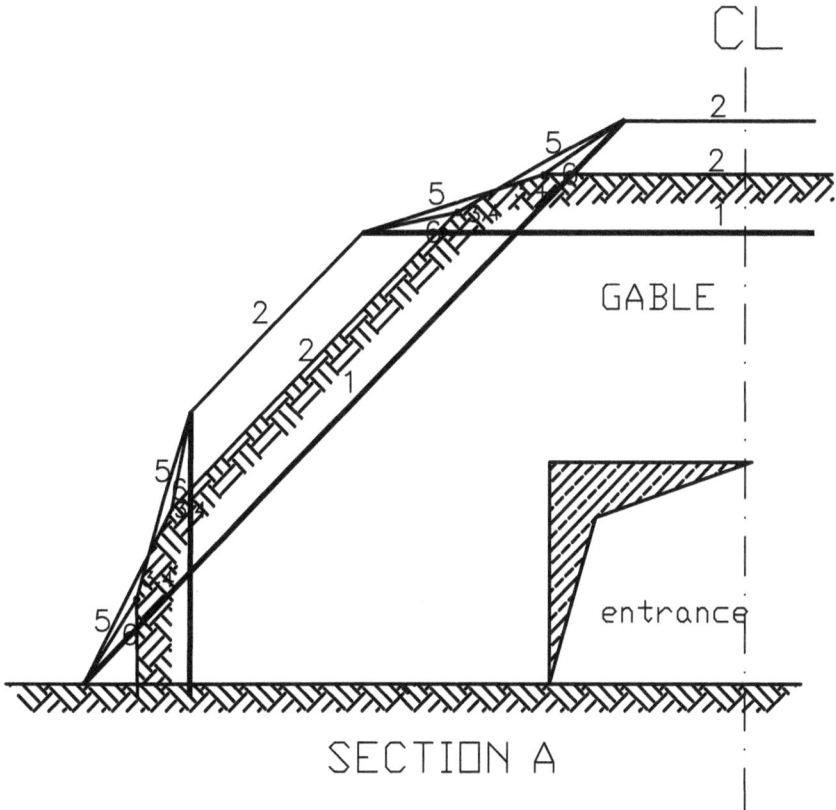

Figure 9.7 Front view of a horizontal tensegrity cylindrical shell.

Chapter 10

Spatial tensegrity arches

The plane tensegrity arch shown in Figure 8.1 can be used to construct the spatial tensegrity arch shown in Figure 10.1.

The spatial tensegrity arch is composed of two systems. System No. 1 consists of members 1, 2, 3′, 4, 5, 6′, 7 and 8. System No. 2, a supporting system,

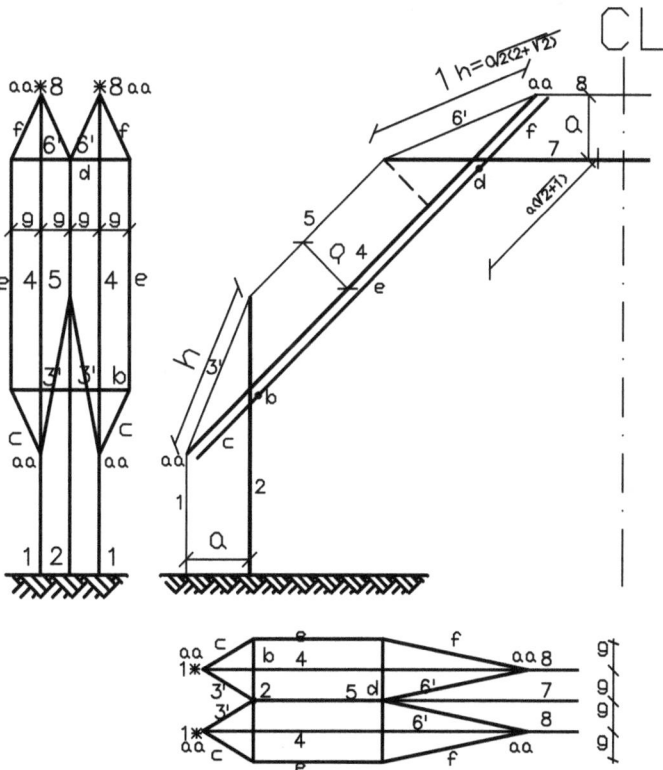

Figure 10.1 Spatial tensegrity arch.

DOI: 10.1201/9781003370093-12

consists of two bars 4 of system No. 1 and additional two bars b and d and cables c, e and f as shown in Figure 10.1.

The forces applied to system No. 1 by prestressing are similar to the forces in the force diagram shown in Figure 8.1. The forces in bars 2 and 7 and cable 5 are shown in this force diagram. The forces in bar 4 and cables 1 and 8 are half of the magnitudes shown in this force diagram. The force F_3' and F_6' in cables 3' and 6' are as follows:

$$10.1 \quad F_3' = F_6' = F_3 \sqrt{h^2 + g^2} / (2h)$$

Here F_3 is the force in cables 3 and 6 of the two-dimensional arch shown in the force diagram in Figure 8.1, and h shown in Figure 10.1 takes the value as follows:

$$10.2 \quad h = a \sqrt{4 + 2\sqrt{2}}$$

The prestressing forces in cables 3' and 6' apply force H to system No. 2, as shown in Figure 10.2.

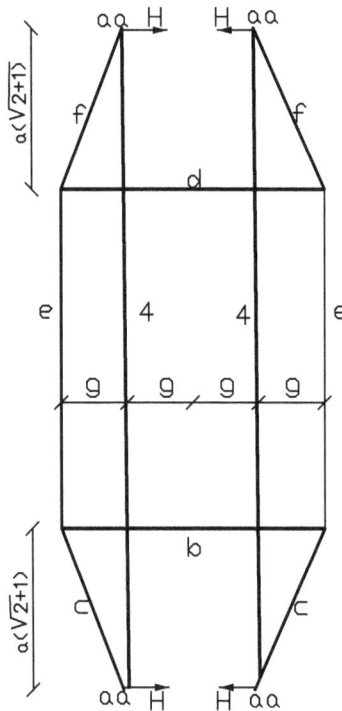

Figure 10.2 Forces applied to system No. 2.

Forces H can be found by using Equation 10.1 and are given by Equation 10.3.

$$10.3 \quad H = gF_3 / (2h)$$

By considering equilibrium at nodes aa, the forces in cables c, e and f and bars b and d can be determined as well as the increase in the compression in bars 4.

The erection of the spatial tensegrity arch can be carried out in two stages. In the first stage, system No. 1 is constructed. This can be done following the method presented in Section 8.1 in which appropriate units with additional structural elements are formed prestressed and placed by using scaffolding in the required locations. In the second stage, system No. 2 is constructed. By connecting it to system No. 1 and prestressing, the additional structural elements used to form the appropriate units can be removed as well as the scaffolding.

Chapter 11

Configuration of primitive tensegrity structures

The primitive tensegrity structures comprises tensegrity plane arches and tensegrity plane rings.

The tensegrity plane arch shown in Figure 8.1 can be used to form two spatial arches shown in Figures 11.1 and 11.2.

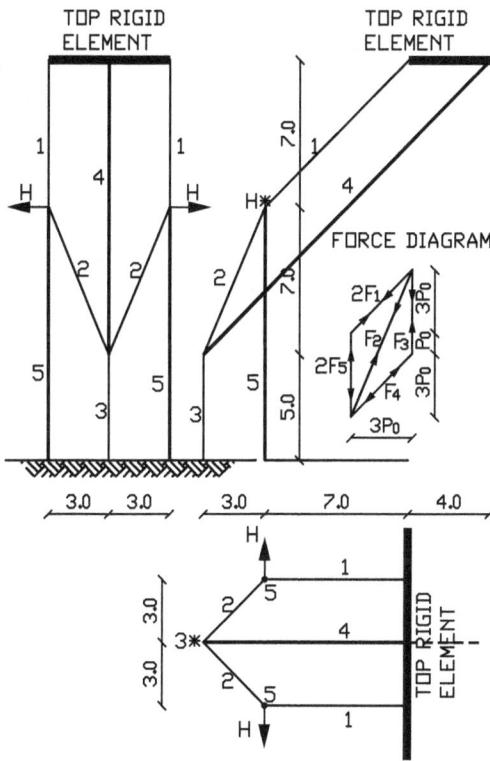

Figure 11.1 Spatial tensegrity arch type A.

DOI: 10.1201/9781003370093-13

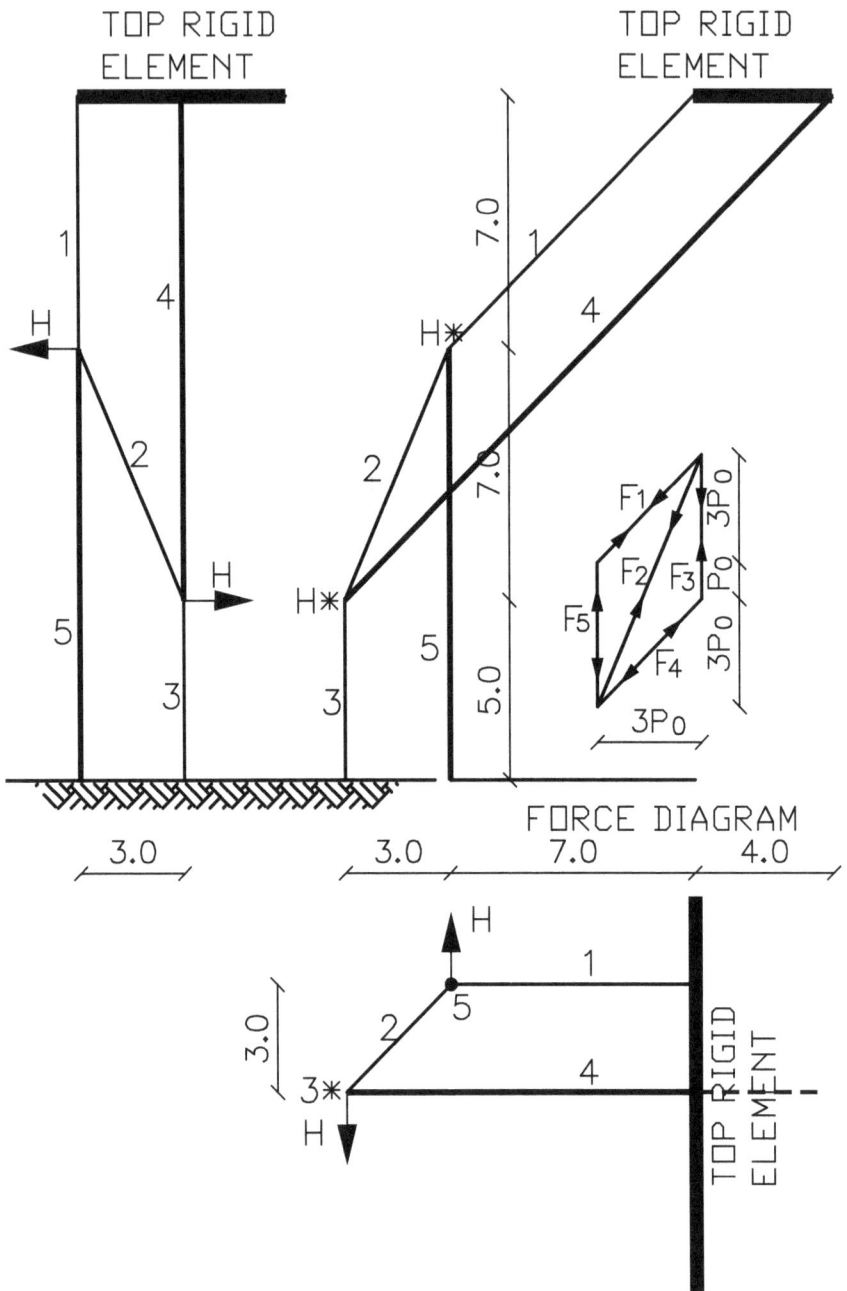

TOP RIGID
ELEMENT

TOP RIGID
ELEMENT

H

1

4

2

5

3

H

7.0

7.0

5.0

1

4

H*

2

5

3

H*

F1

F2 F3

F5

F4

3P0

P0

3P0

3P0

FORCE DIAGRAM

3.0

3.0

7.0

4.0

H

1

5

2

4

3*

H

3.0

TOP RIGID
ELEMENT

Figure 11.2 Spatial tensegrity arch type B.

From the force diagram presented in Figures 11.1 and 11.2, it can be seen that horizontal forces H of $3P_0$ are required to keep these two spatial tensegrity arches in equilibrium. The tensegrity plane ring shown in Figure 5.1 is used to form the one shown in Figure 11.3.

To avoid entanglement between bars 1' and bars 2', bar 1' is divided into two elements with a gap between them. Bars 2' are passing through this gap. This method can be adopted in all cases where members of a tensegrity structure intersect. The force diagram in Figure 11.3 shows the forces in the tensegrity ring elements when cable 4 is prestressed to $3P_0$.

The spatial tensegrity arch type A shown in Figure 11.1 and the tensegrity ring shown in Figure 11.3 are used to form the primitive tensegrity dome type A shown in Figures 11.4 and 11.5 and the primitive tensegrity cooling tower type A shown in Figures 11.6 and 11.7.

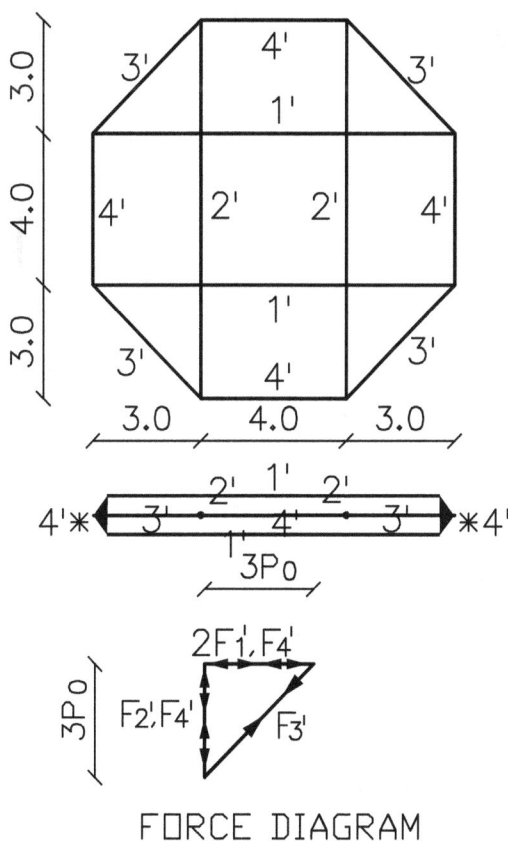

FORCE DIAGRAM

Figure 11.3 Tensegrity ring.

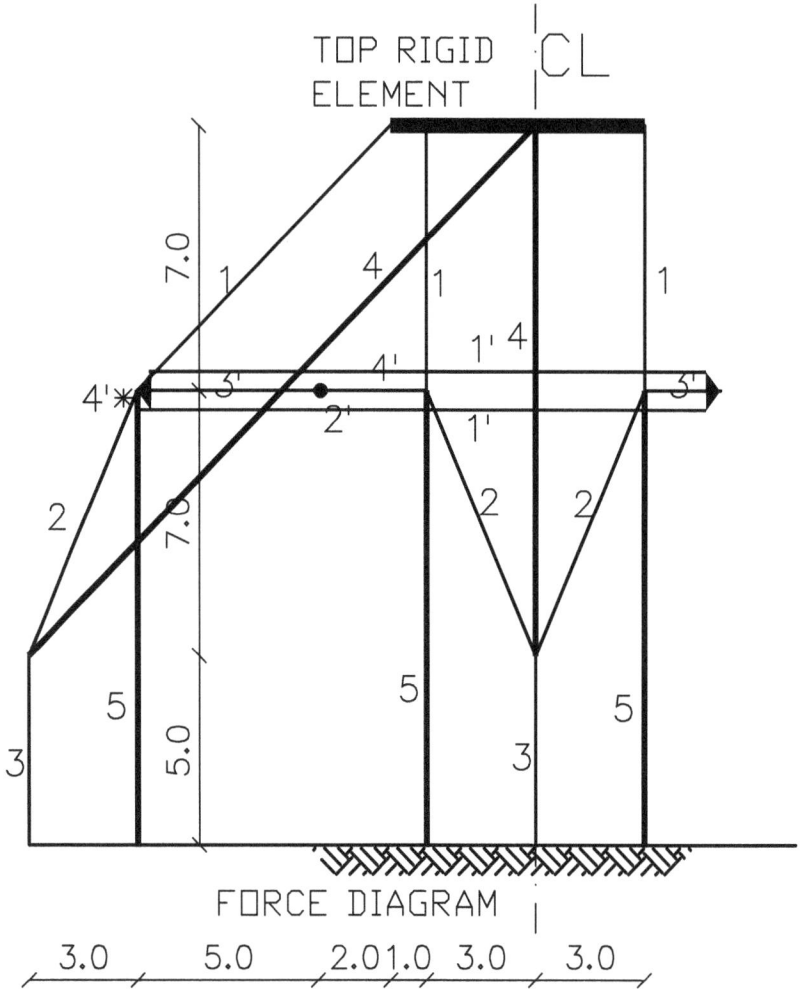

Figure 11.4 Side view of a primitive tensegrity dome type A.

The spatial tensegrity arch type B shown in Figure 11.2 and the tensegrity ring shown in Figure 11.3 are used to form the primitive tensegrity dome type B shown in Figures 11.8 and 11.9.

It is easy to use the spatial tensegrity arch type B to form a primitive type B cooling tower.

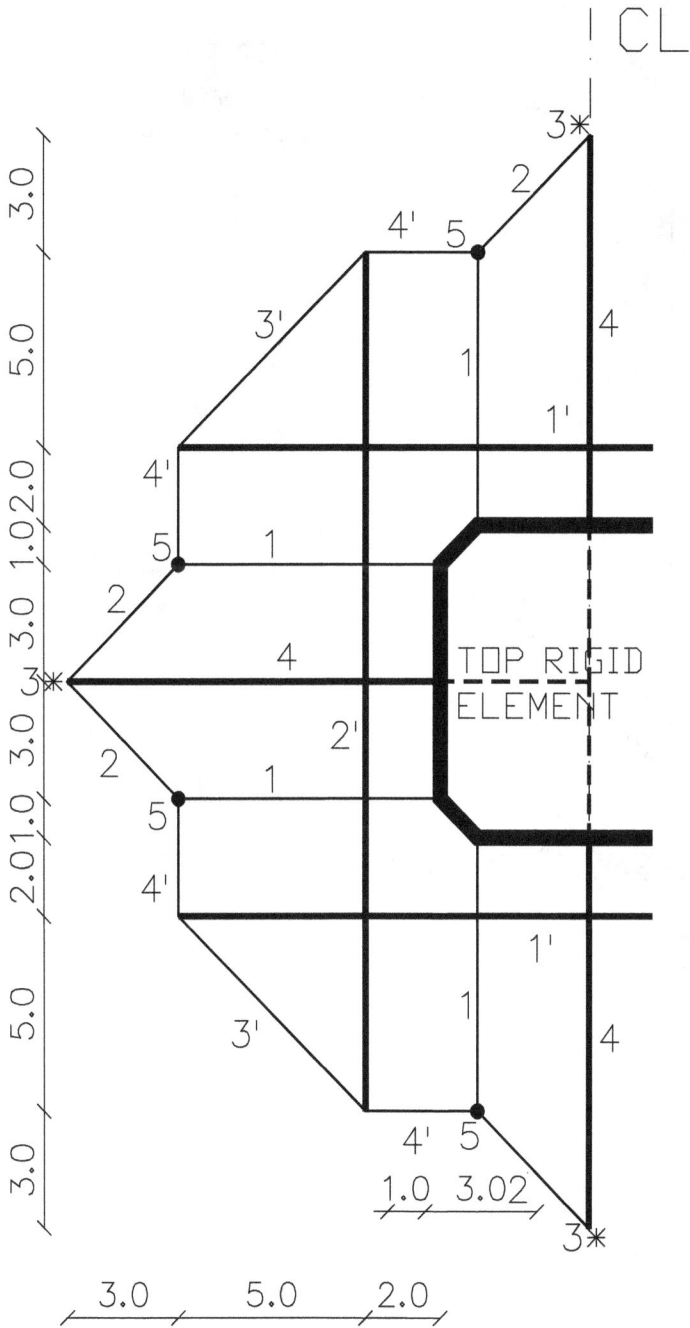

Figure 11.5 Plan of a primitive tensegrity dome type A.

TOP RIGID CL
ELEMENT

5.0

7.0

3

5

2

1'

4'* 3' 4'
2' 1'

7.0

4

1

1'

1 4 1

4.0 7.0 2.0 1.0 1.0 3.0 3.0

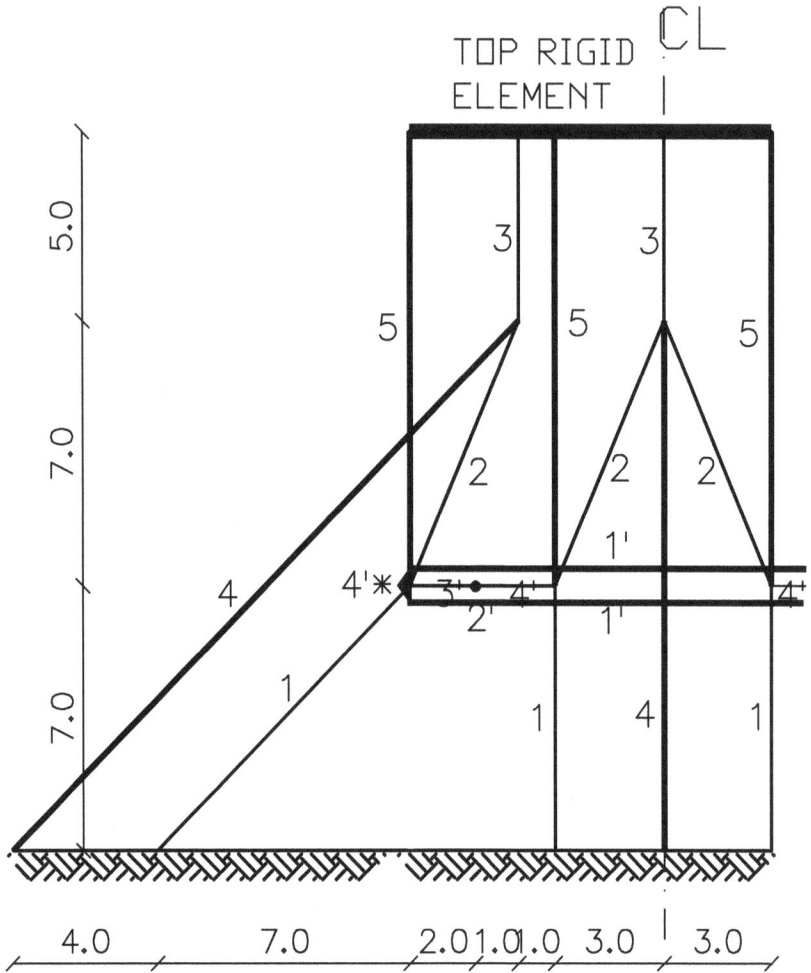

Figure 11.6 Side view of a primitive tensegrity cooling tower type A.

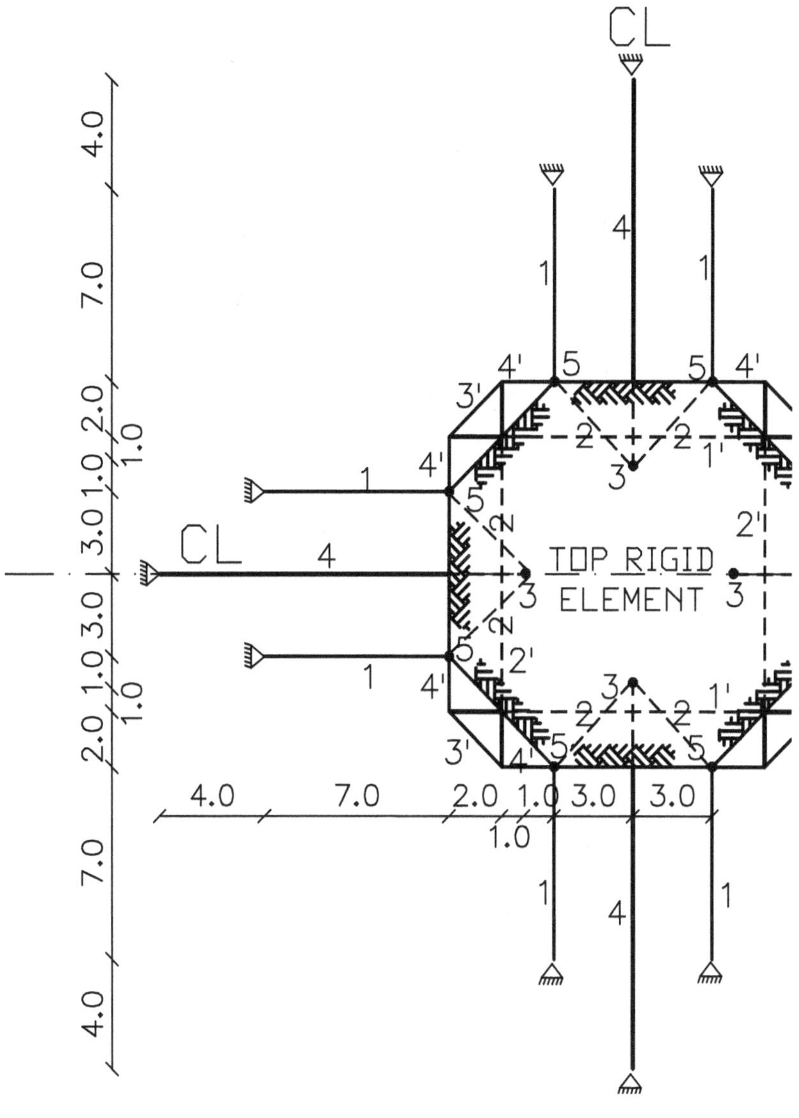

Figure 11.7 Plane of a primitive tensegrity cooling tower type A.

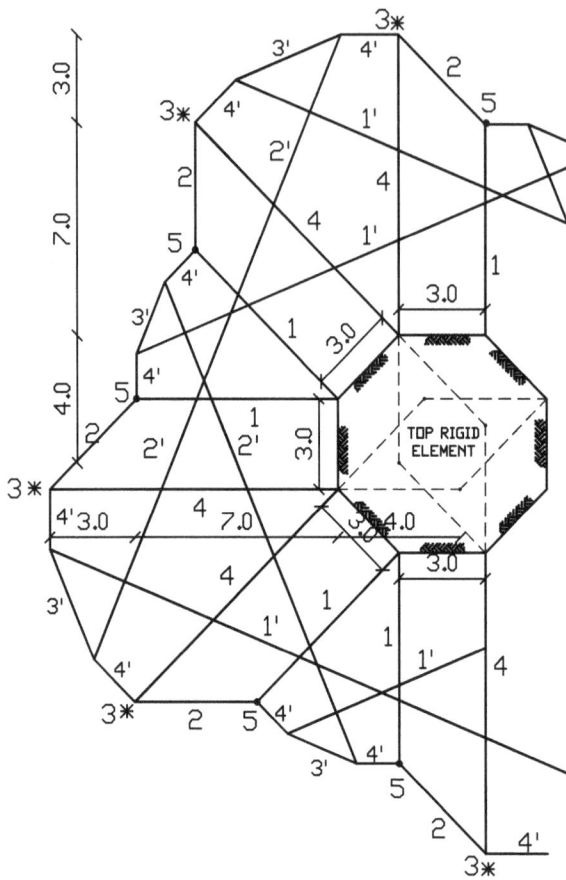

Figure 11.8 Plan of a primitive tensegrity dome type B.

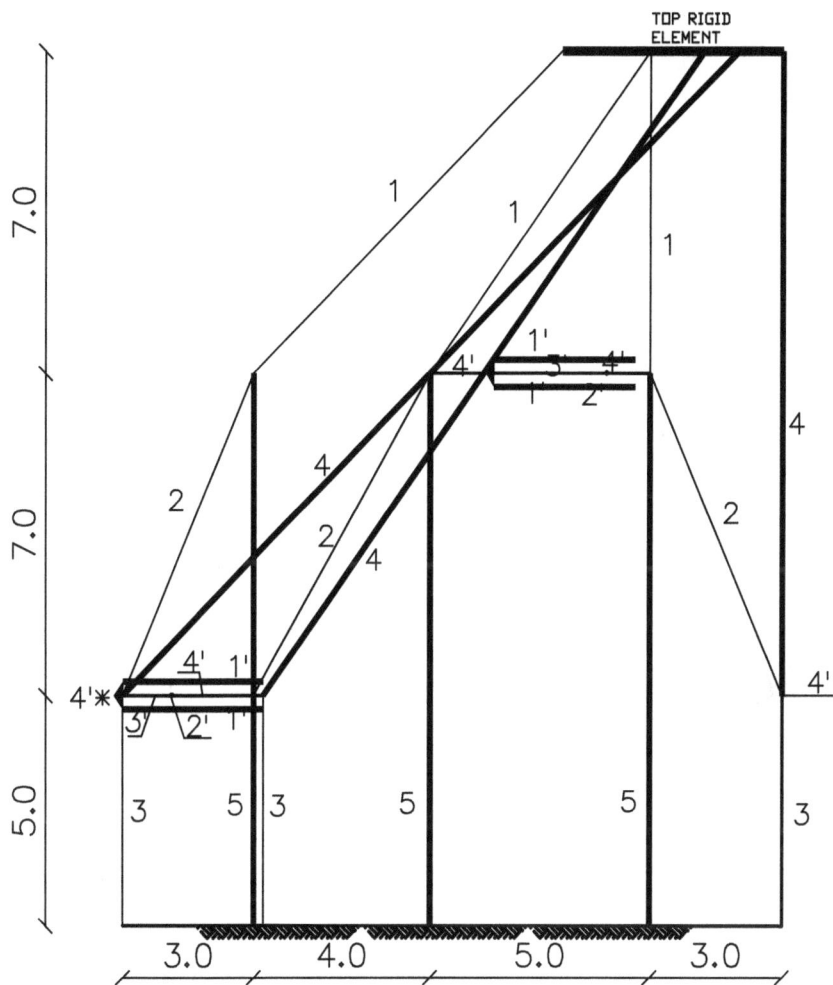

Figure 11.9 Side view of a primitive tensegrity dome type B.

Chapter 12

Slim tensegrity domes

12.1 SLIM VERTICAL TENSEGRITY DOMES

The slim vertical tensegrity dome consists of spatial tensegrity arches connected by cables, as shown in Figures 12.1 and 12.2. In this example, the

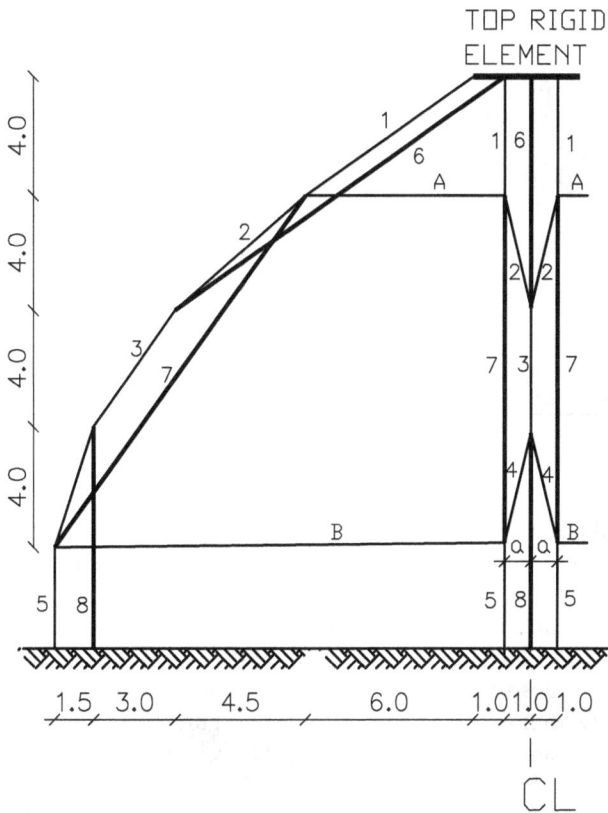

Figure 12.1 Side views of the slim vertical tensegrity dome.

DOI: 10.1201/9781003370093-14

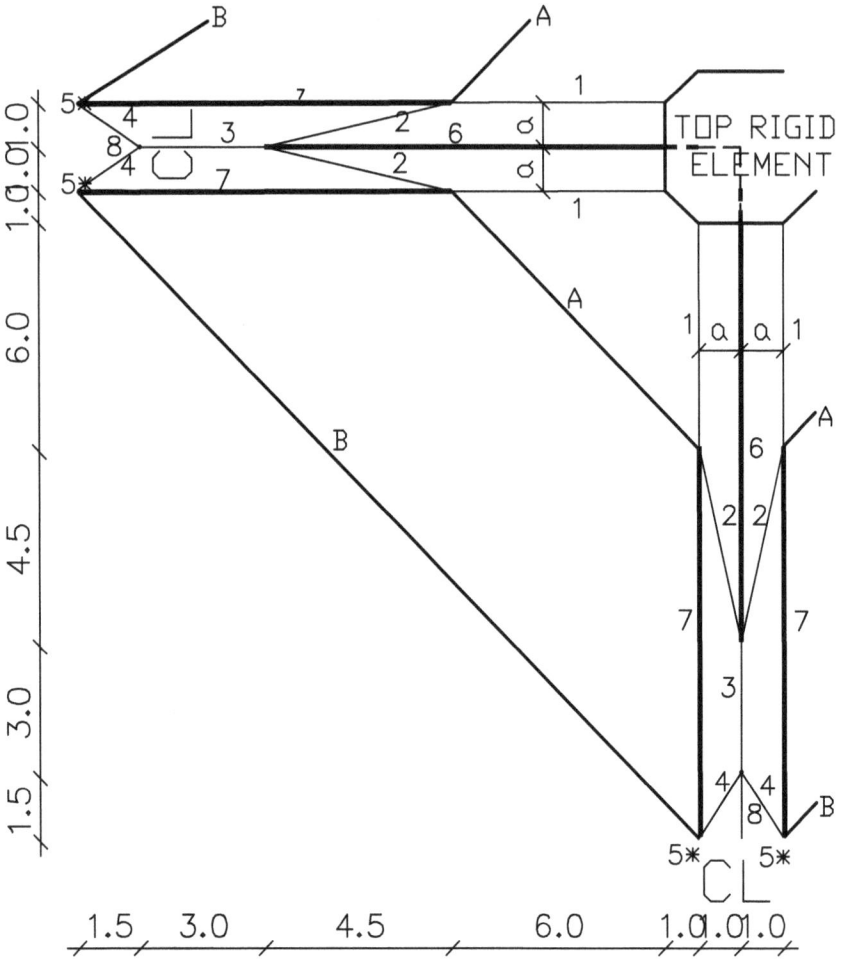

Figure 12.2 Plan of the slim vertical tensegrity dome.

slim vertical tensegrity dome is built of four spatial tensegrity arches. Cables A and B connect these spatial tensegrity arches and provide the forces required to keep these tensegrity arches in equilibrium.

The configuration of the tensegrity arch is based on the suspension cable loaded by equal point loads L1 = L2 = L3 = L4 shown in Figure 12.3.

By using the suspension cable shown in Figure 12.3, the configuration of the tensegrity arch takes the form shown in Figure 12.4.

The prestressing forces of the tensegrity arch shown in Figure 12.4 are displayed in the force diagram. It should be noted that since there are two elements to cables 1, 2, 4, 5 and bar 7, the force in them is half of what is shown in the force diagram.

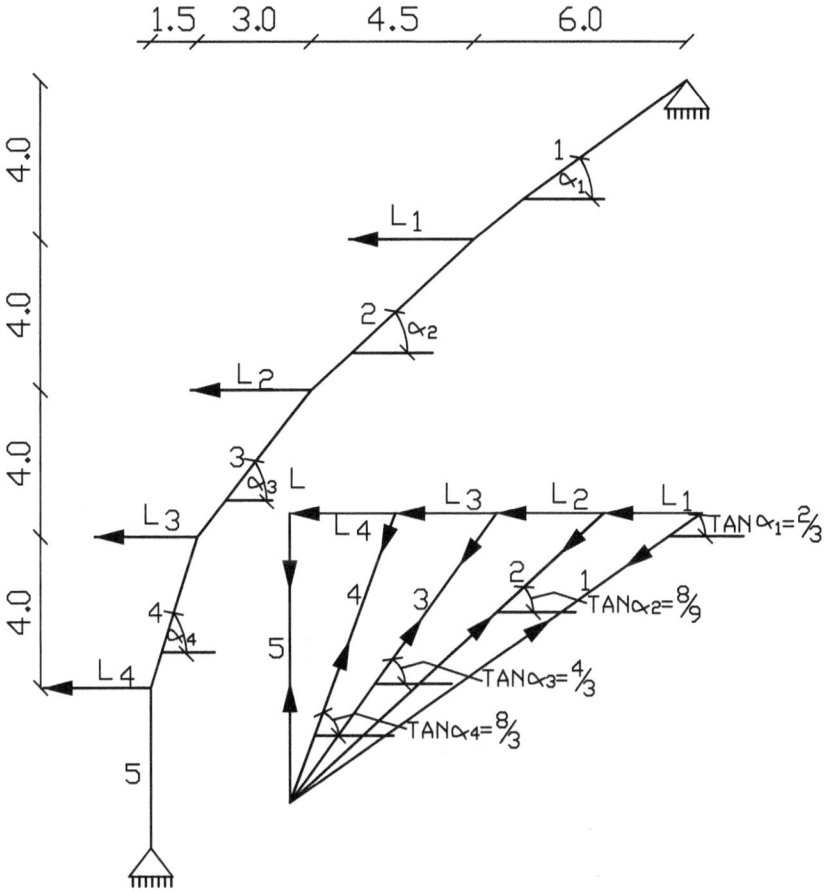

Figure 12.3 Suspension cable loaded by equal point loads.

In the case where the tensegrity arch is part of the slim vertical tensegrity dome, the supporting cables A and B apply the tangential forces A^T and B^T to the tensegrity arches as, shown in Figure 12.4. Equilibrium at the appropriate nodes aa and bb implies that the forces A^T and B^T take the following form:

$$12.1 \quad A^T = aF^2_z / l^2_z, \quad B^T = F^4_z a / l^4_z$$

Because there are two cables 2 and 4 in each tensegrity arch,

$$12.2 \quad l^2_z = l^4_z = l_0 \text{ and } F^2_z = F^4_z = P_0$$

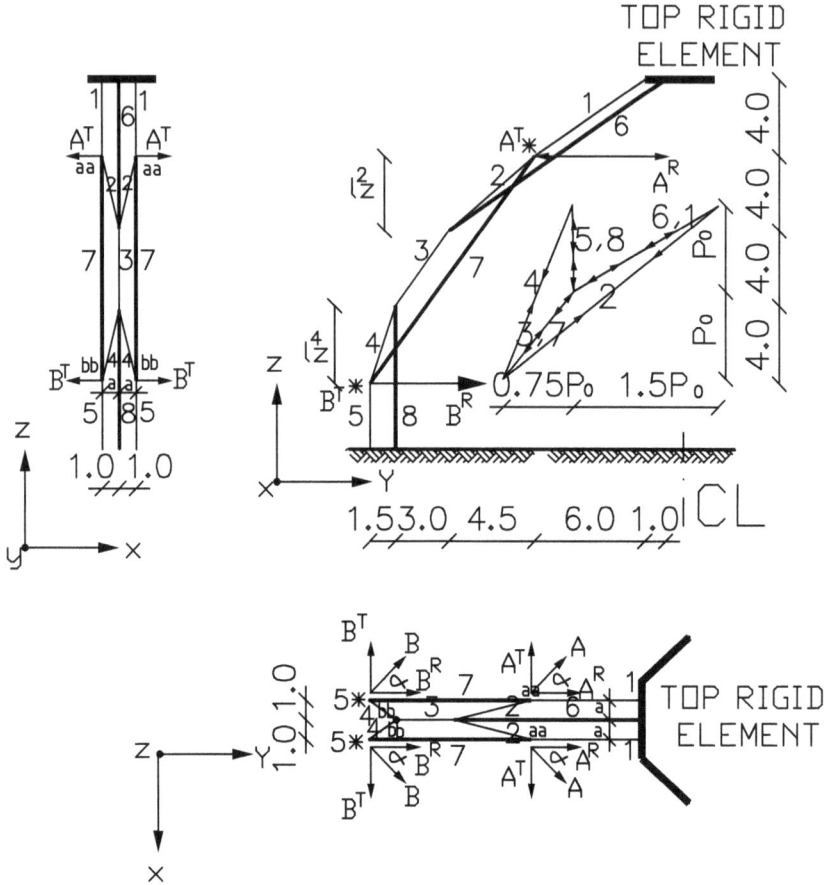

Figure 12.4 Tensegrity arch based on the suspension cable shown in Figure 12.3.

Equation 12.1 takes the following form:

$$12.3 \quad A^T = B^T = aP_0 / l_0$$

The supporting cables A and B apply also radial horizontal forces A^R and B^R to the tensegrity arch as shown in Figure 12.4.

$$12.4 \quad A^R = B^R = aP_0 \tan \alpha / l_0$$

Because the configuration of this tensegrity arch is based on the suspension cable shown in Figure 12.3, the equal loads A^R and B^R shown in Figure 12.4 are "fitted load". The tensegrity arch can sustain then in its prestressed configuration. In the case where cable 2 is prestressed so that F^2_z is equal to P_0, A^R and B^R change only the forces in elements 1, 7 and 5, as shown in Figure 12.5.

Forces A^R and B^R increase the compression in bar 7 and reduce the tension in cables 1 and 5.

The compression ΔF^1_z induced to the two cables 1 in the z-direction can be found by using Figure 12.5:

$$12.5 \quad \Delta F^1_z = 4aP_0 \tan \alpha / (3l_0)$$

The tension in cable 1 in z-direction is as follows:

$$12.6 \quad F^1_z = F^5_z = P_0/2 - 4aP_0 \tan \alpha /(3l_0)$$

To keep cable 1 in tension, "a" should be as follows:

$$12.7 \quad a < 3l_0 /(8 \tan \alpha)$$

As the number of tensegrity arches used to form the slim vertical tensegrity dome increases, $\tan \alpha$ is reduced and so the magnitude of "a" can be increased.

In the example of the slim vertical tensegrity dome presented in Figures 12.1 and 2.2, l_0 is equal to 4.0 and $\tan \alpha$ is equal to 1.0. In this case, a = 1.0 satisfies Equation 12.7.

Because of the reduction in the tension in cable 1, a vertical uplift force is applied to the top rigid element.

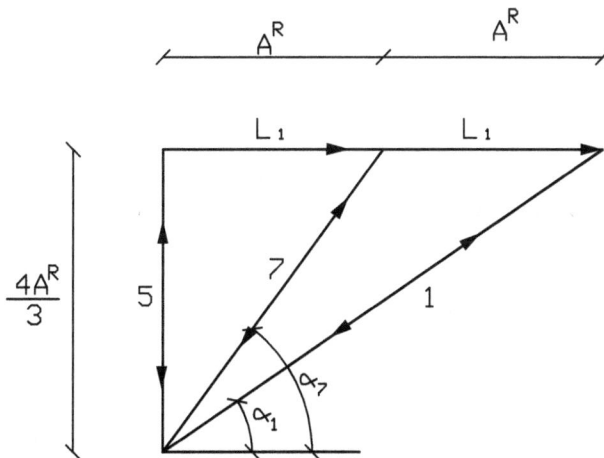

Figure 12.5 Forces induced to members 1, 7 and 5 due to A^R and B^R.

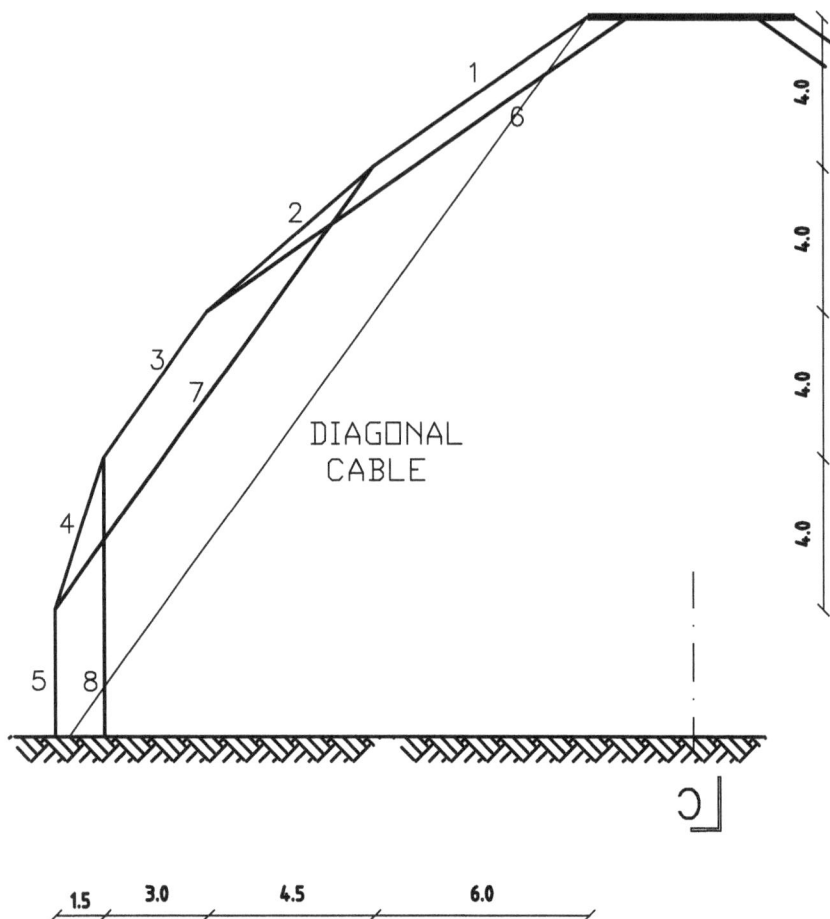

Figure 12.6 Proposed diagonal cables to sustain the vertical cable sustaining the uplift forces.

By using Equation 12.5, the uplift force acting on the top rigid element, UP, due to one tensegrity arch, considering the fact that each tensegrity arch has two cables 1, is as follows:

$$12.8 \quad \text{UP} = 8aP_0 \tan \alpha / (3l_0)$$

The vertical uplift force applied to the top rigid element can be carried out by diagonal cables as shown in Figure 12.6.

Because the slim vertical tensegrity dome is appropriately prestressed, it is a feasible slim vertical tensegrity dome and can be constructed in this proposed configuration.

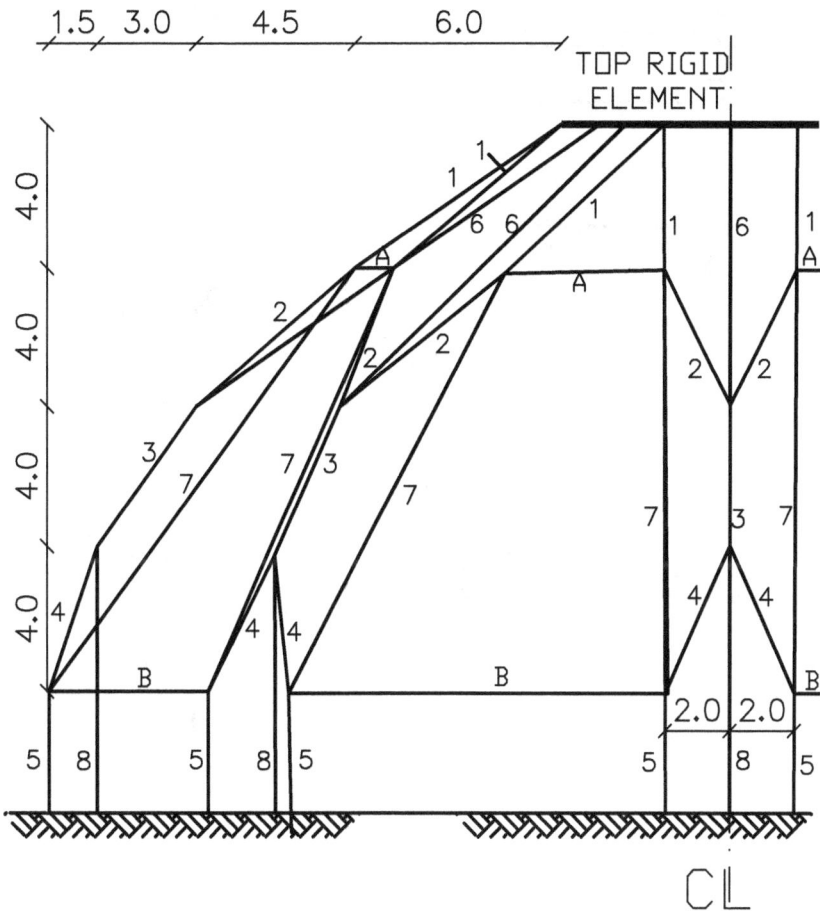

Figure 12.7 Side view of a slim vertical tensegrity dome with eight tensegrity arches.

An example of a slim vertical tensegrity dome with eight tensegrity arches is shown in Figures 12.6 and 12.7. The analysis of this slim vertical tensegrity dome can be carried out following the analysis of the slim vertical tensegrity dome with four tensegrity arches described before. It can be seen that in this case because α is equal to 22.5°, "a" can be larger than the "a" in the case of the slim vertical tensegrity dome shown in Figures 12.1 and 12.2. In this case, a = 2.0 satisfies Equation 12.7. This slim tensegrity dome with eight tensegrity arches is shown in Figures 12.7 and 12.8.

The slim vertical tensegrity dome can be erected following the methods used in the case of the level tensegrity vault presented in Section 8.1. In the case of the slim vertical tensegrity dome, there are three units A, B and C shown in Figure 12.9.

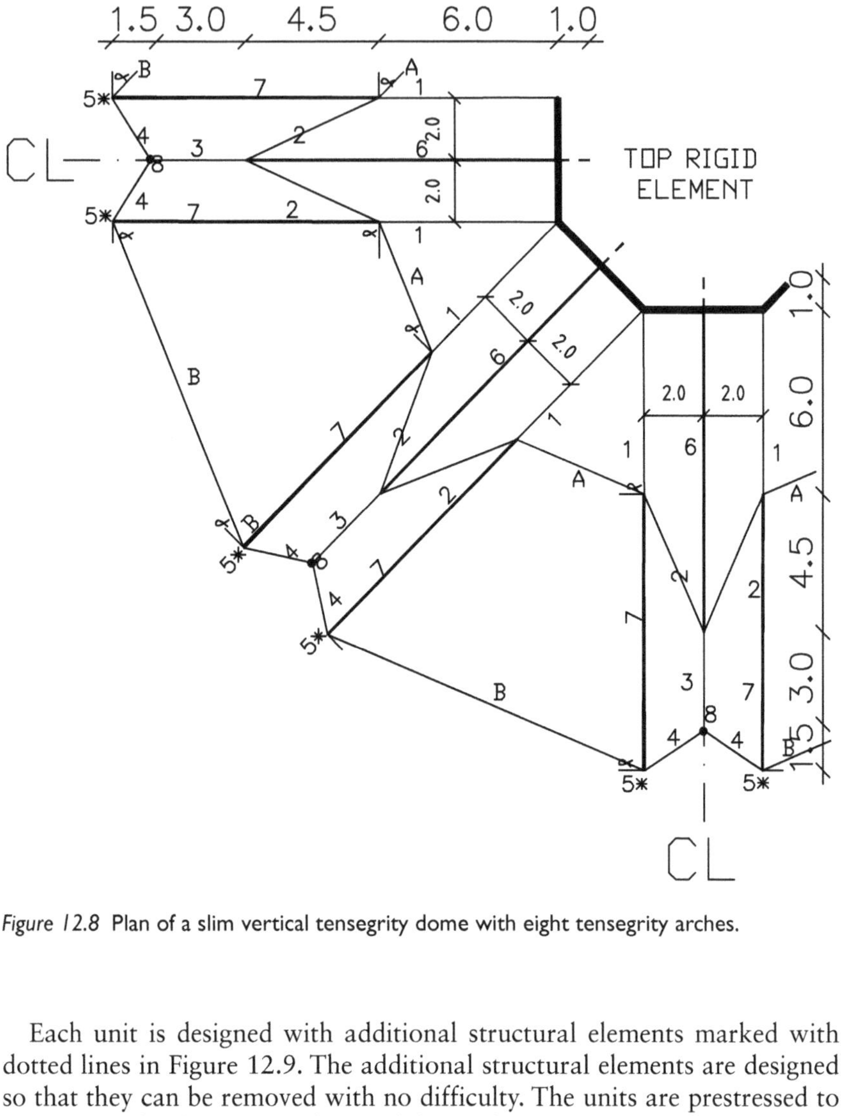

Figure 12.8 Plan of a slim vertical tensegrity dome with eight tensegrity arches.

Each unit is designed with additional structural elements marked with dotted lines in Figure 12.9. The additional structural elements are designed so that they can be removed with no difficulty. The units are prestressed to a minimum level to ensure their stability only. After the units are placed by using appropriate scaffoldings, in their location the identical nodes should be joined together. Cables A and B are installed as well as the diagonal cables shown in Figure 12.6, also these cables are prestressed to the minimum level to ensure that all cables are under tension. Then the slim vertical tensegrity dome is prestressed to the extent that it is possible to remove the additional structural elements of units A, B and C. After removing the additional structural elements, the slim vertical tensegrity dome can be prestressed to the designed level.

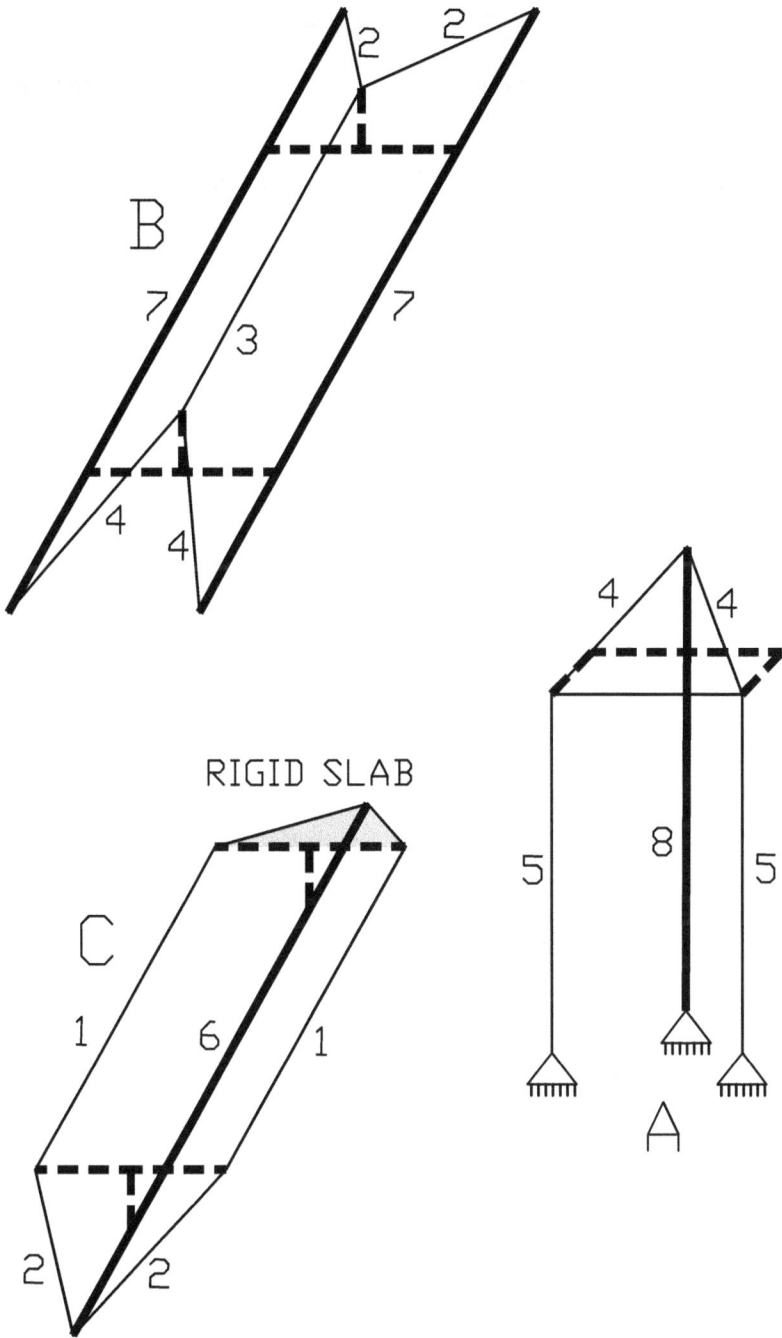

Figure 12.9 Basic units for the construction of the slim vertical dome.

12.2 SLIM RECLINING TENSEGRITY DOMES

The slim reclining tensegrity dome is designed following the footsteps of the design of the slim vertical tensegrity dome. In the case of the slim reclining tensegrity dome, the spatial tensegrity arches span horizontally between two gables placed on both sides of the slim reclining tensegrity dome. In this case, the connecting cables A and B are vertical.

The analysis of the forces induced by prestressing to the slim reclining tensegrity dome follows the method used in the analysis of the slim vertical tensegrity dome.

Plan of a slim reclining tensegrity dome is shown in Figure 12.10, a front view in Figure 12.11 and a side view in Figure 12.12.

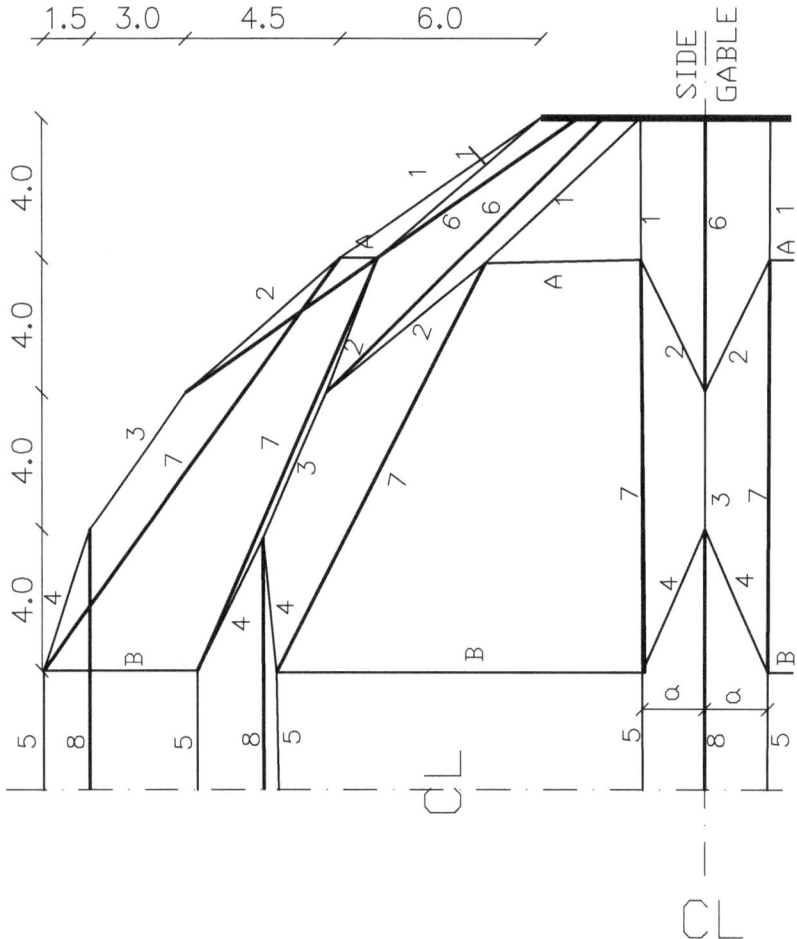

Figure 12.10 Plan of a slim reclining tensegrity dome.

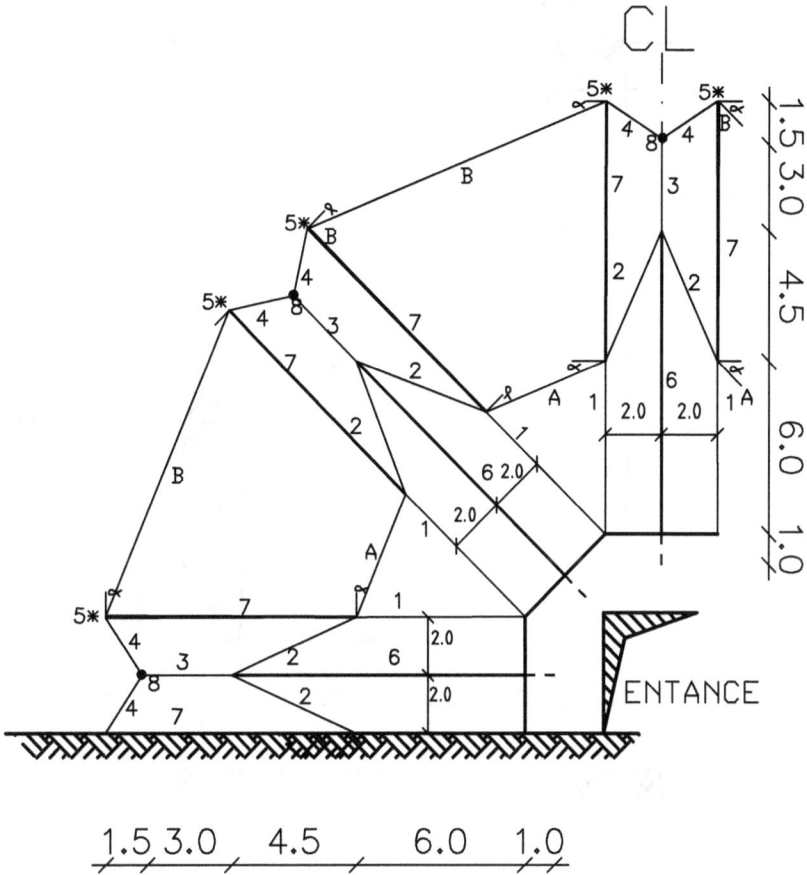

Figure 12.11 Front view of a slim reclining tensegrity dome.

In the case of the slim reclining tensegrity dome, the uplift forces of the slim vertical tensegrity dome applied to the top rigid element are horizontal. So in the case of a slim reclining tensegrity dome, the gables should be designed to sustain this horizontal force.

The slim reclining tensegrity dome can be erected following the method used in the case of the slim vertical dome shown in Section 12.1.

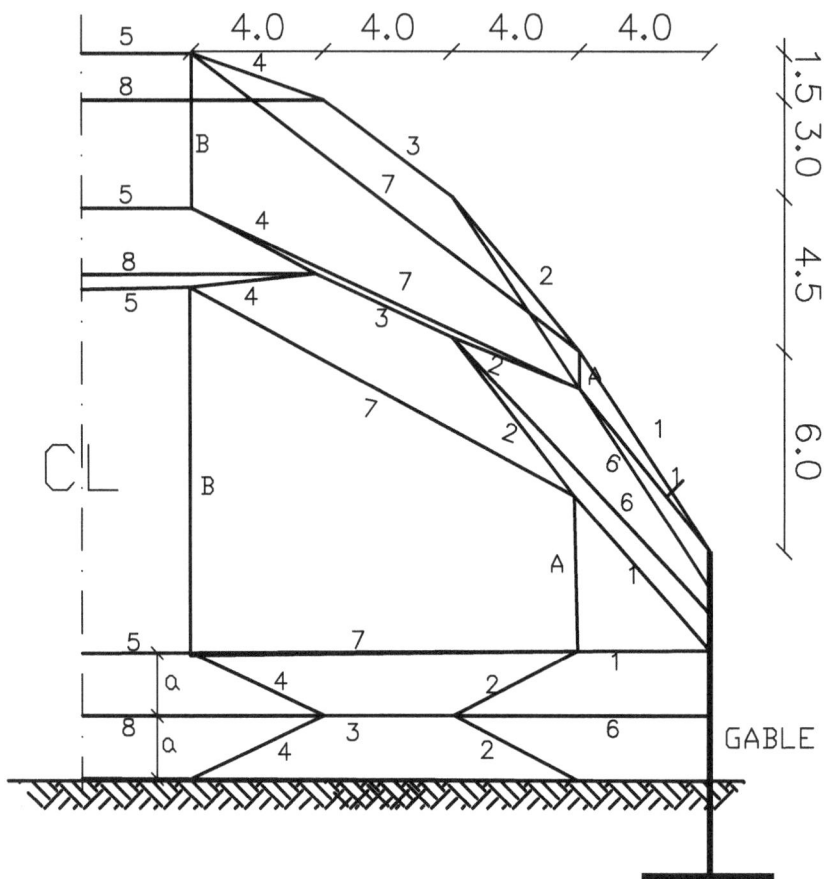

Figure 12.12 Side view of a slim reclining tensegrity dome.

Chapter 13

Tensegrity caps

13.1 TALL TENSEGRITY CAPS

The tall tensegrity cap is a crosswise arrangement of spatial tensegrity arches. The case of a tall tensegrity cap composed of four spatial tensegrity arches is shown in Figures 13.1 and 13.2. Each of the four spatial tensegrity arches is marked. The spatial tensegrity arches are connected to each other by connecting cables shown by dotted lines in Figures 13.1 and 13.2.

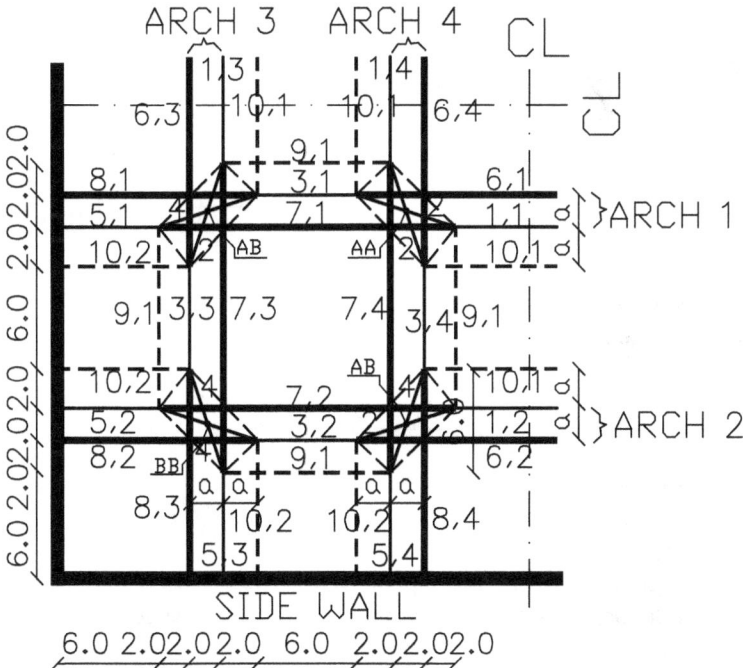

Figure 13.1 Plan of a tall tensegrity cap.

DOI: 10.1201/9781003370093-15

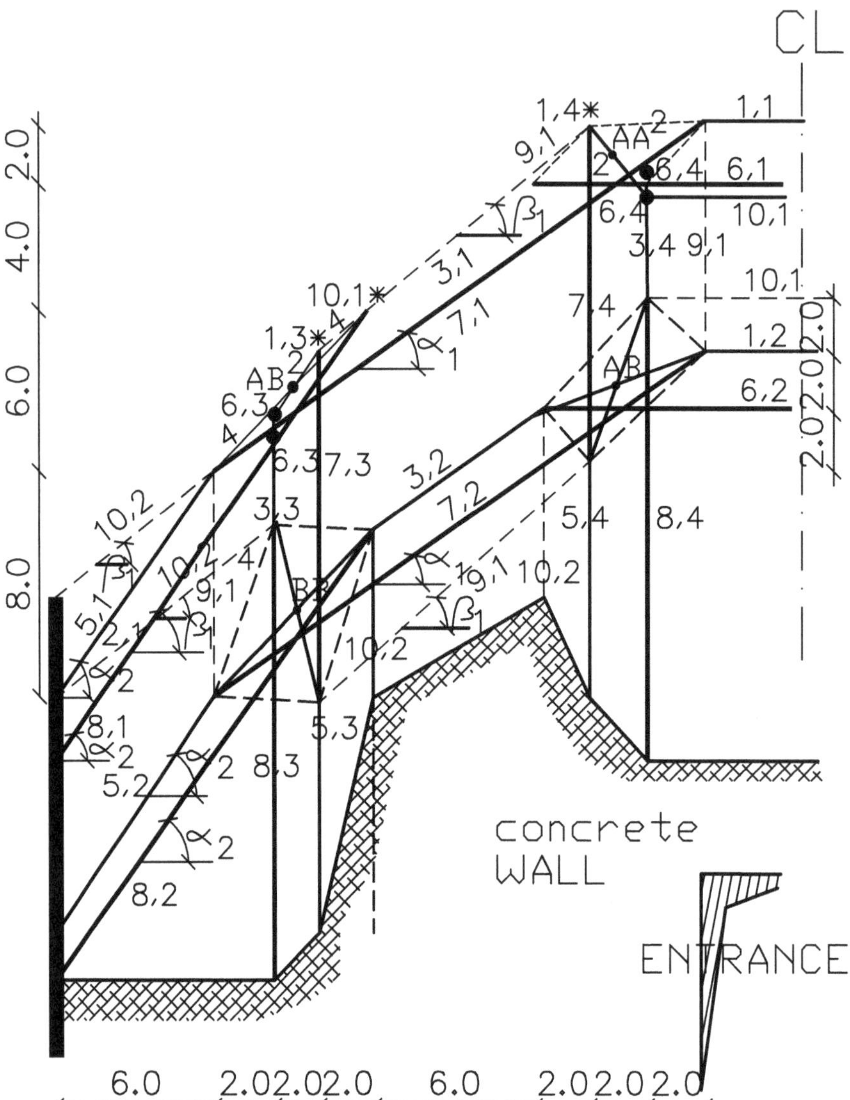

Figure 13.2 Side view of a tall tensegrity cap.

The configuration of the four spatial tensegrity arches is based on a suspension cable loaded with equal vertical load at equal spacing as shown in Figure 13.3.

The spatial tensegrity arch, based on this suspension cable, takes the form shown in Figure 13.4

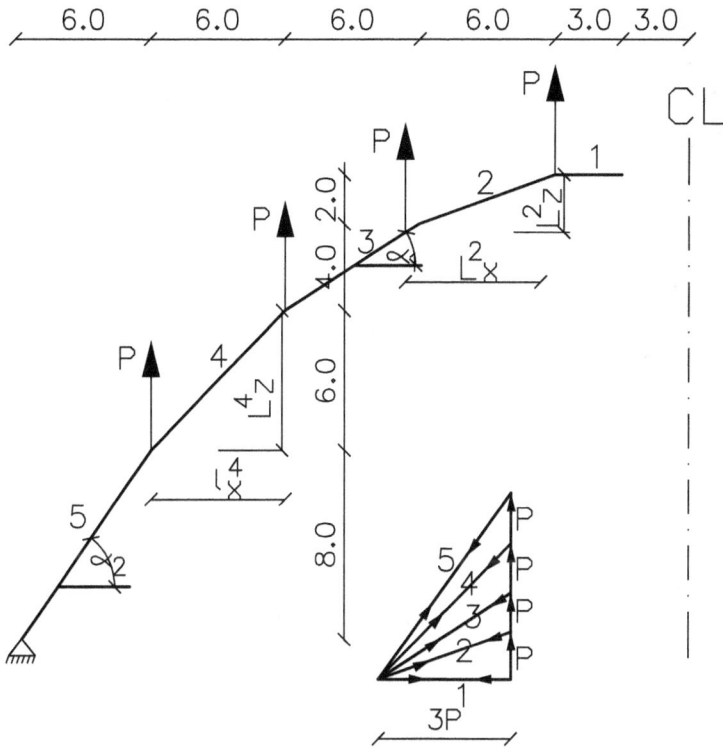

Figure 13.3 Suspension cable used to form the configuration of the tensegrity arch.

Points A and B shown in Figure 13.4 are at the middle of cables 2 and 4, respectively. The crosswise arrangement of the spatial tensegrity arches is dictated by the fact that two spatial tensegrity arches that cross each other share the appropriate points A or B accordingly at the same common location as shown in Figures 13.1 and 13.2.

To avoid entanglement of bars 6,4 and 6,1, bar 6,4 is split into two elements with a gap between them to let bar 6,4 to pass through. Also bar 6,3 is split into two elements to avoid entanglement with bar 7,1

The forces induced by prestressing the spatial tensegrity arch shown in Figure 13.4 are shown in Figure 13.5.

The following relationship between the absolute value of the forces can be observed:

$$13.1 \quad F^2_x = F^4_x = 18P_0$$

$$13.2 \quad \begin{aligned} F^1_x &= F^3_x = F^5_x = 9P_0 \\ F^6_x &= F^7_x = F^8_x = F^4_x = -9P_0 \end{aligned}$$

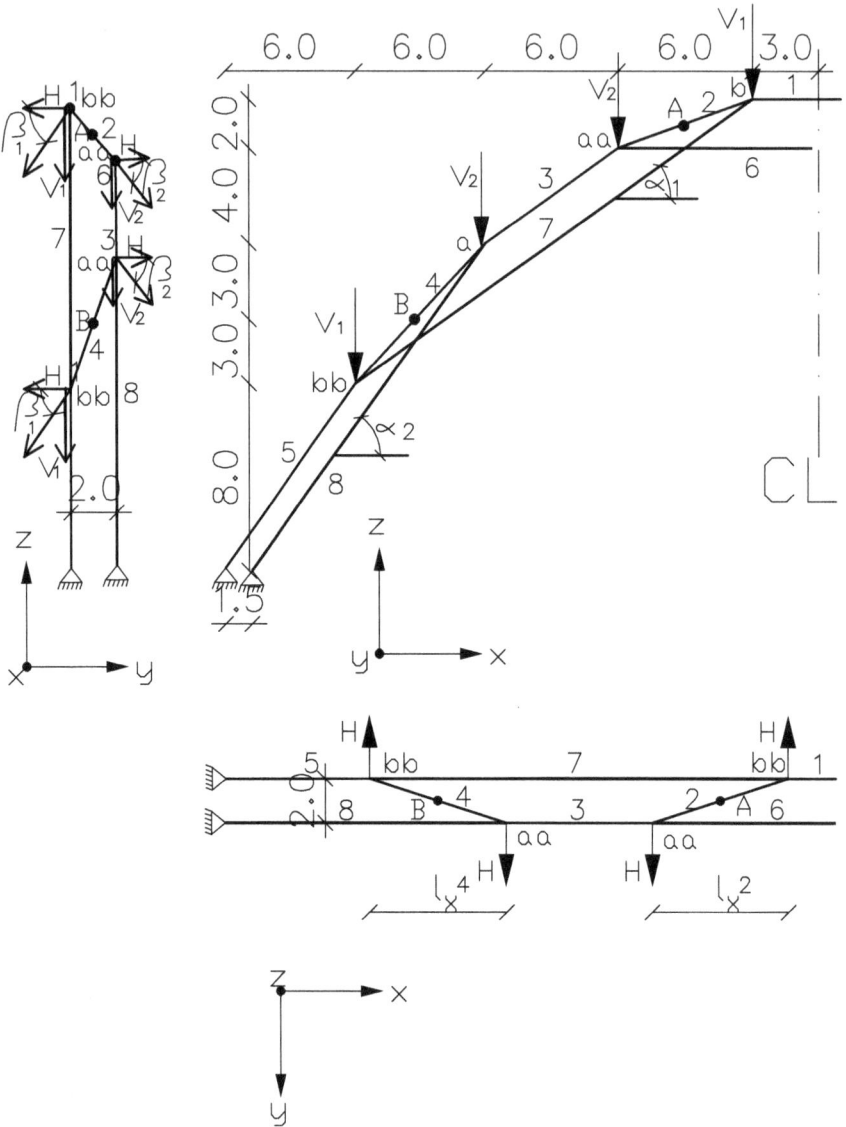

Figure 13.4 Spatial tensegrity arch based on the suspension cable shown in Figure 13.3.

Here, F^i_x is the component of force i in the X-direction.

Horizontal force H given by Equation 13.3 is required to maintain equilibrium at nodes aa and bb of the spatial tensegrity arch shown in Figure 13.4.

$$13.3 \quad H = F^2_x a / L^2_x = F^4_x a / L^4_x = 18aPo / L^2_x$$

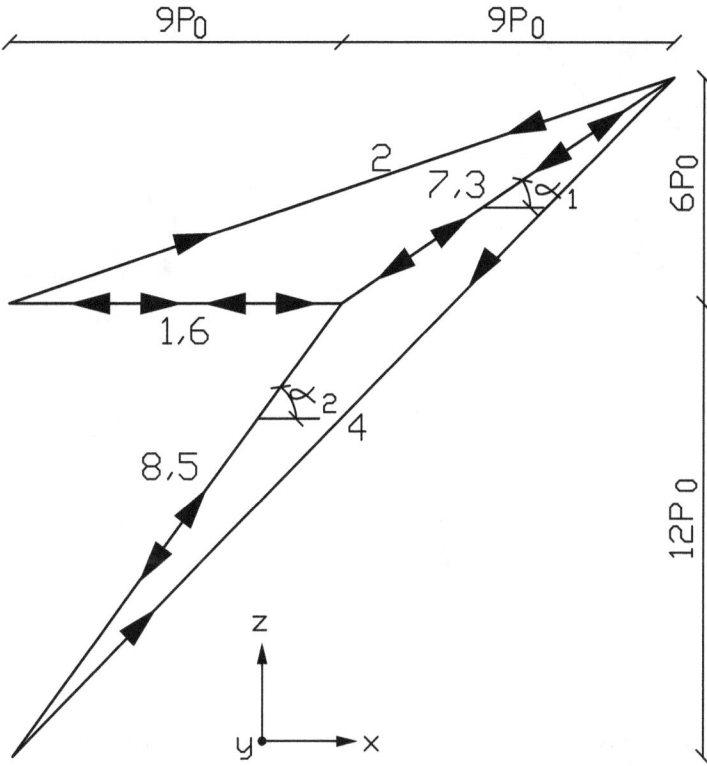

Figure 13.5 Prestressing forces of the tensegrity arch shown in Figure 13.4.

Here, L^2_x and L^4_x are the lengths of cables 2 and 4 in the X-direction.

Forces H are applied to the spatial tensegrity arches by the connecting cables 9,1, 10,1 and 10,2 marked by dotted lines in Figures 13.1 and 13.2. In the case where the inclination of the connecting cables at nodes aa are $\beta 1$ and at node bb are $\beta 2$, the supporting cables apply the vertical forces V1 and V2 to the spatial tensegrity arch given by Equations 13.4 and 13.5:

$$13.4 \quad \textbf{V1} = H\tan\beta 1 \ = 18aP_0\tan\beta 1 / L^2_x$$

$$13.5 \quad \textbf{V2} = H\tan\beta 2 \ = 18aP_0\tan\beta 2 / L^2_x$$

Because the configuration of the tensegrity arch is based on a suspension cable loaded by equal loads, **V1** and **V2** are "fitted loads". The spatial tensegrity arch can sustain these loads in its prestressed configuration. **V1** loads change the forces in elements 6, 3 and 8, **V2** loads change the forces

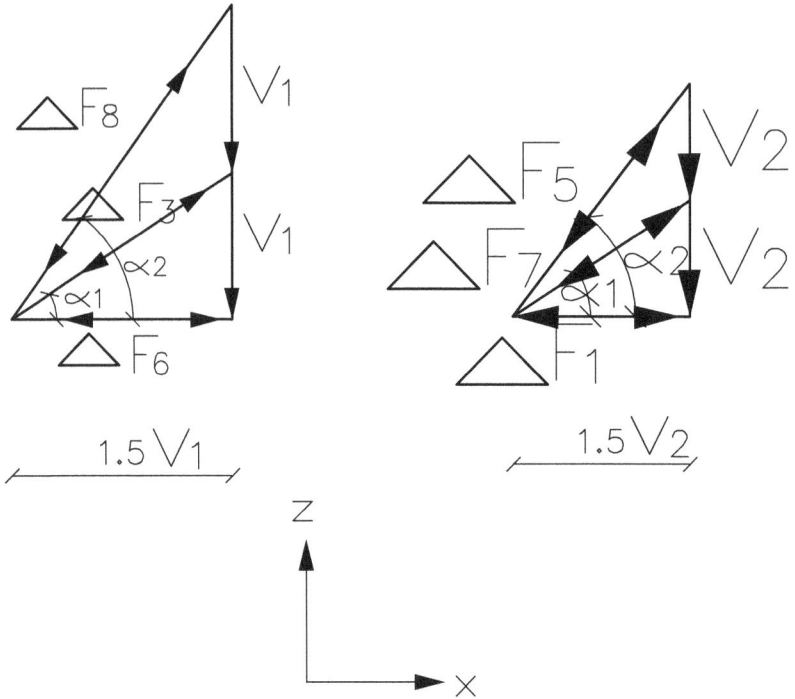

Figure 13.6 Change of forces in the spatial tensegrity arch elements due to the supporting cables.

in elements 1, 5 and 7 of the spatial tensegrity arches. This change can be determined by considering the force diagram shown in Figure 13.6.

Since tan $\alpha_1 = 2/3$, the compression induced to the spatial tensegrity arch members is as follows:

$$13.6 \quad \begin{aligned} \Delta F^1_x &= \Delta F^7_x = \Delta F^5_x = 1.5\mathbf{V}2 = -27P_0 \tan \beta1 / L^2_x \\ \Delta F^3_x &= \Delta F^6_x = \Delta F^8_x = 1.5\mathbf{V}1 = -27P_0 \tan \beta2 / L^2_x \end{aligned}$$

By considering the change in the forces in the spatial tensegrity arch members given by Equation 13.6, tension is maintained in cables 1 and 5 when Equation 13.7 is satisfied.

$$13.7 \quad 9P_0 > 27aP_0 \tan \beta1 / L^2_x$$

Tension in cable 3 is maintained when Equation 13.8 is satisfied.

$$13.8 \quad 9P_0 > 27aP_0 \tan \beta2 / L^2_x$$

Equations 13.7 and 13.8 imply:

$$13.9 \quad \begin{array}{l} a < L^2_x / (3 \tan \beta 1) \\ a < L^2_x / (3 \tan \beta 2) \end{array}$$

In the case where $L^2_x = 6.0$, Equation 13.9 takes the following form:

$$13.10 \quad \begin{array}{l} a < 2 / \tan \beta 1 \\ a < 2 / \tan \beta 2 \end{array}$$

In the case of the tall tensegrity cap shown in Figures 13.1 and 13.2, the maximum inclination of the connecting cables, $\beta 1$, takes the value of $\tan \beta 1 = 0.8$. In this case, Equation 13.10 implies that "a" should satisfy Equation 13.11.

$$13.11 \quad a < 2.5$$

In the case of the tall tensegrity cap shown in Figures 13.1 and 13.2, "a" is assumed to be 2.0.

In some cases **V1** and **V2** induce tension to some bars. Because the bars can sustain tension, these cases are of minor importance.

The forces induced into the elements of the tall tensegrity cap can be determined in the following way. By considering the force diagram shown in Figure 12.5, the initial forces in numbers (1,1), (3,1), (5,1), (6,1), (7,1), (8,1), (1,2), (3,2), (5,2), (6,2), (7,2), (8,2), (1,3), (3,3), (5,3), (6,3), (7,3), (8,3), (1,4), (3,4), (5,4), (6,4), (7,4) and (8,4) can be determined.

By using Equation 12.3, the horizontal component of the forces in the supporting cables (9,1), (10,1) and (10,2) can be determined.

By considering the inclination of the supporting cables, the forces induced by the prestressing can be determined. The inclination of cables 9,1 and 10,2 is $\beta 1$, the inclination of cable 10,1 is zero and the inclination of 10,2 is also $\beta 1$.

By considering the inclination of the supporting cables (9,1), (10,1), (10,2), Equations 13.4, 13.5 and the force diagram shown in Figure 13.6, the change in the force acting in the tensegrity arch elements due to the supporting cables can be determined. It can be seen that because cables (10,1) are horizontal, it does not cause to a change in the forces in numbers (3,1), (6,1) (8,1), (3,4), (6,4) and (8,4). Because of the inclination of cables (9,1), it induces compression into cables (1,1), (5,1), (1,4) and (5,4) and compression to bars (7,1) and (7,4). Cable (9,1) applies an uplift load to arch 3 and arch 2; it increases the tension in cables (1,2), (5,2), (1,3); and (5,3) and it reduces the compression in bar (7,2) and (7,3). The forces in cables (10,2) induce compression to cables (3,2) and (3,3) and increases the compressions to bars (6,2), (8,2), (6,3) and (8,3).

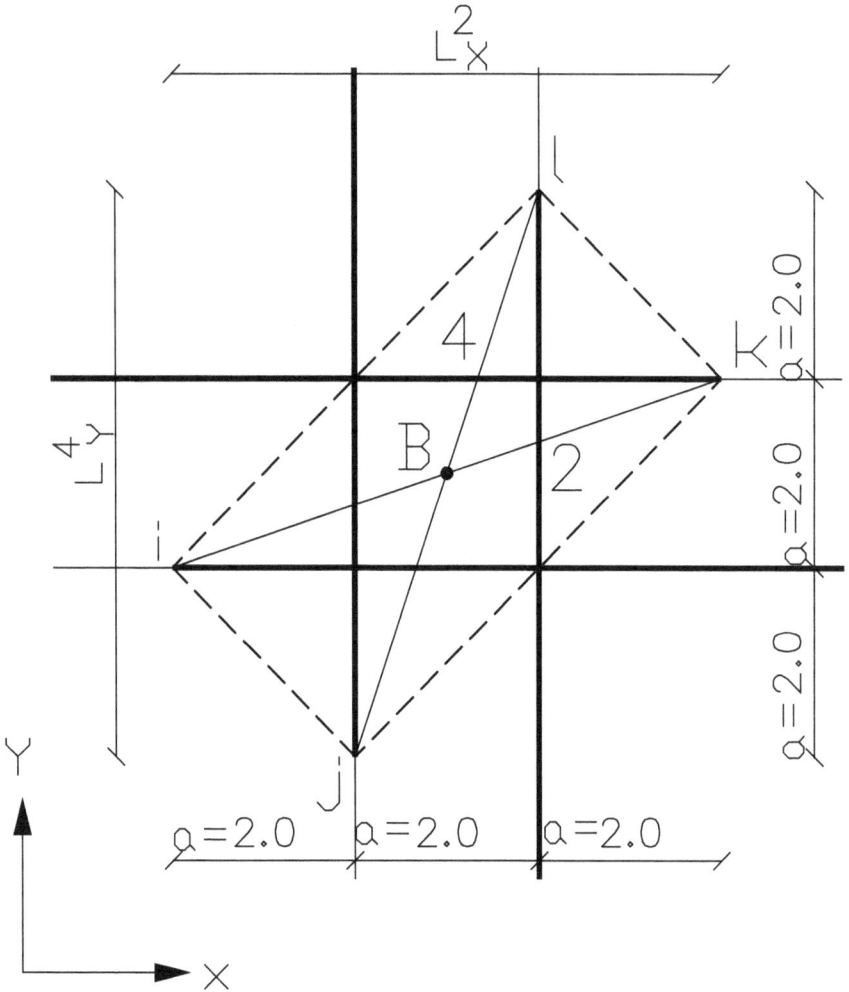

Figure 13.7 Replacing cables 2 and 4 by four cables.

Cables 2 and 4 are straight and intersect at a common point at the middle as shown in Figures 13.7, so points i, j, k and l form a plane as shown in Figure 13.7.

It can be seen that Equation 1.9 of Chapter 1 is satisfied.

$$13.12 \quad \begin{array}{l} ((F^2_x)^2 + (F^2_z)^2 + H^2) / ((F^4_x)^2 + (F^4_z)^2 + H^2) = \\ ((L^2_x)^2 + (L^2_z)^2 + a^2) / ((L^4_x)^2 + (L^4_z)^2 + a^2) \end{array}$$

So cables 2 and 4 can be replaced by four cables L_{ij}, L_{ik}, L_{kl} and L_{ik} shown by the dotted lines in Figures 13.7, 13.1 and 13.2.

Also when two cables 2 or two cables 4 intersect Equation 1.9 is obviously satisfied and they can be replaced by four cables as shown by the dotted lines in Figures 13.1 and 13.2.

The fact that the tall tensegrity dome can be appropriately prestressed in the given configuration implies that this tall tensegrity cap is a feasible tensegrity structure.

13.2 SHALLOW TENSEGRITY CAPS

The configuration of the shallow tensegrity cap is based on the configuration of the tall tensegrity cap shown in Figures 13.1 and 13.2.

It comprises the top section of the tall tensegrity cap and it is supported by a horizontal ring. The plan of the shallow tensegrity cap is shown in Figure 13.8 and a side view in Figure 13.9.

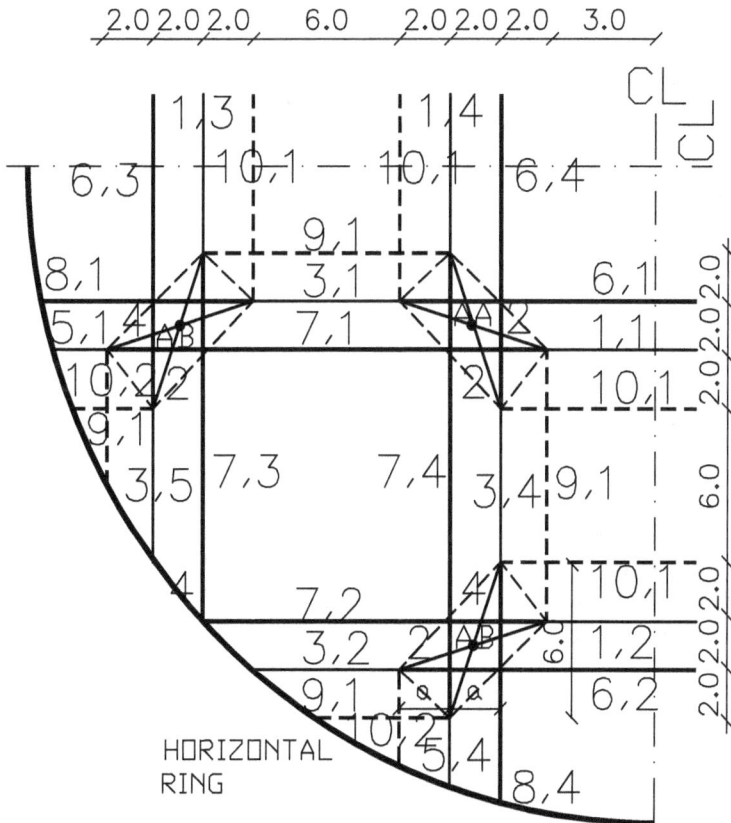

Figure 13.8 Plan of the shallow tensegrity cap.

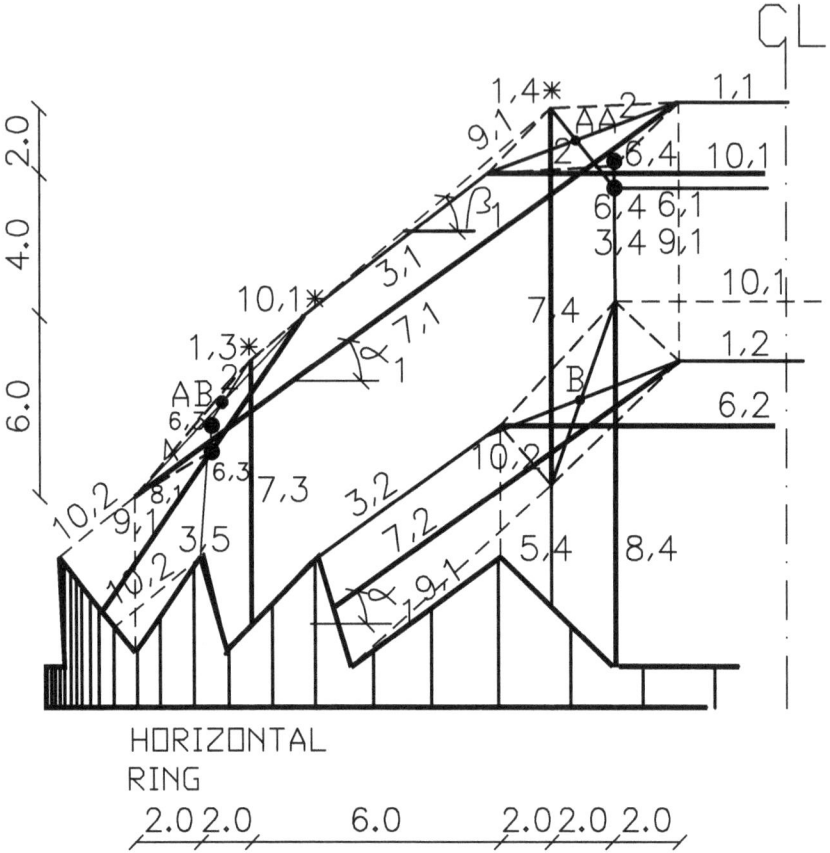

Figure 13.9 Side view of the shallow tensegrity cap.

The analysis of the forces induced to the members of the shallow tensegrity cap can be carried out by using the method described in the case of the tall tensegrity cap discussed in Section 13.1.

The fact that the shallow tensegrity cap can be appropriately prestressed in this configuration indicates that this shallow tensegrity cap is a feasible tensegrity structure.

Chapter 14

Tensegrity domes

14.1 UPRIGHT TENSEGRITY DOMES

The upright tensegrity dome comprises tensegrity arches and tensegrity rings. The rings connect the tensegrity arches to each other.

A side view and a plan of an upright tensegrity dome with four tensegrity arches are shown in Figures 14.1 and 14.2. The tensegrity arches are connected by cables A.

The configuration of the tensegrity arches shown in Figure 14.3 is based on the suspension cable shown in Figure 14.4.

The forces induced by prestressing the tensegrity arches are shown in Figure 14.5.

The magnitude of the vertical components of the force in cables 2 and 4 are as follows: F^2_z and F^4_z are equal to $2P_0$ in tension. In cables 1, 3 and 5, F^1_z, F^3_z and F^5_z are P_0 tension; in bars 6, 7 and 8, F^6_z, F^7_z, F^8_z are P_0 compression.

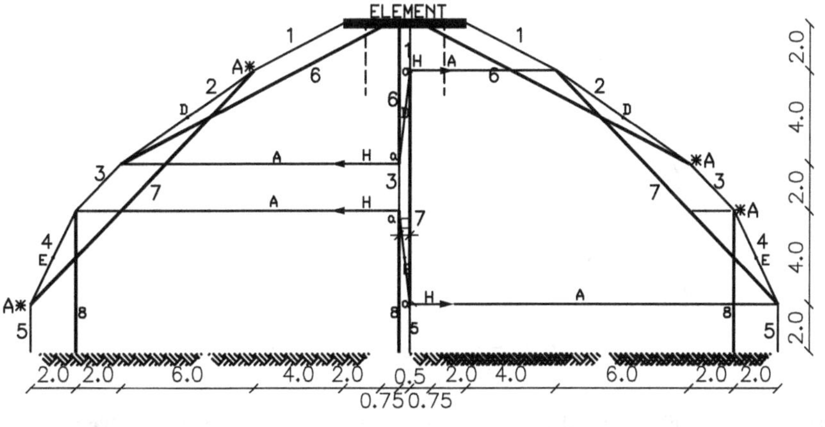

Figure 14.1 Side view of an upright tensegrity dome of four tensegrity arches.

DOI: 10.1201/9781003370093-16

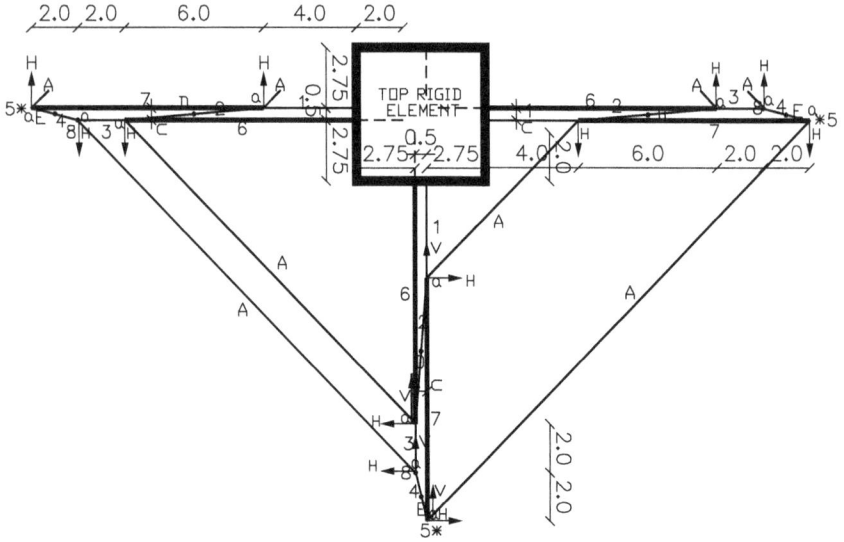

Figure 14.2 Plan of an upright tensegrity dome of four tensegrity arches.

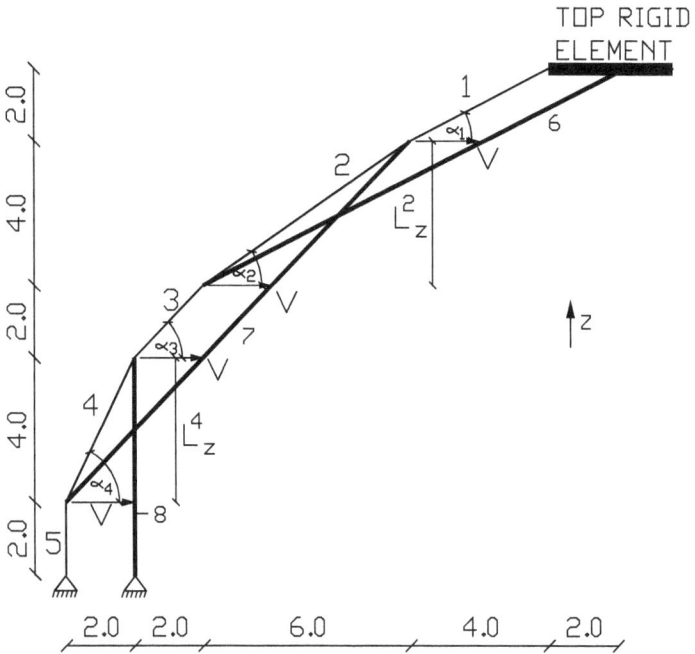

Figure 14.3 Configuration of the tensegrity arch based on the suspension cable shown in Figure 14.4.

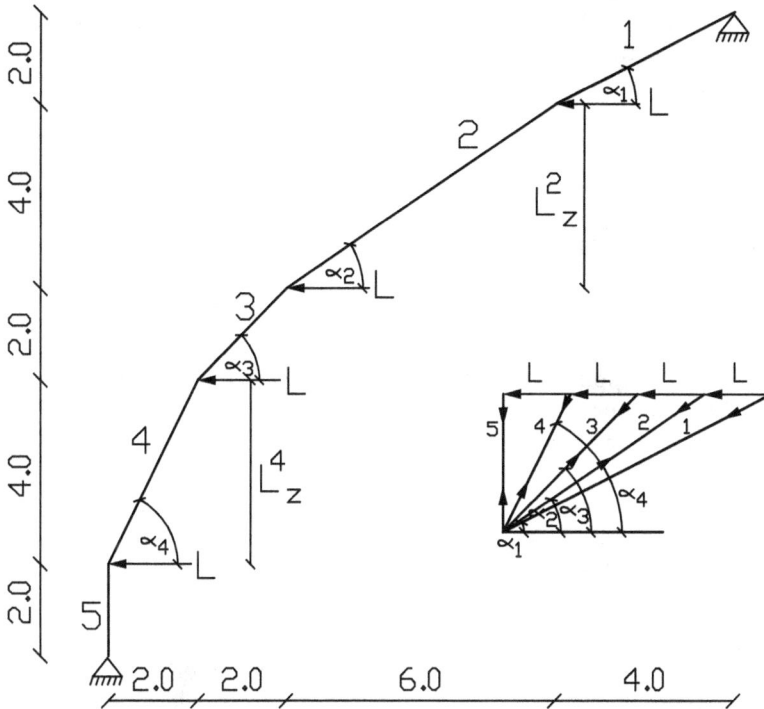

Figure 14.4 Configuration of the suspension cable.

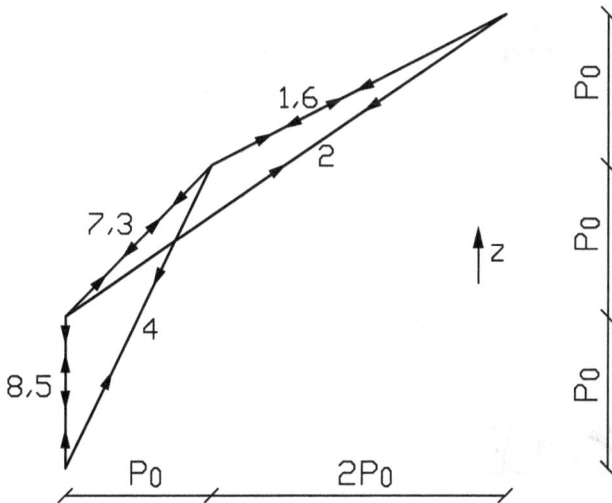

Figure 14.5 Force diagram of the prestressing forces of the tensegrity arch shown in Figure 14.3.

The tensegrity arches of the upright tensegrity dome are spatial structures. Force H associated with cables 2 and 4 is required to maintain equilibrium at nodes a shown in Figures 14.1 and 14.2.

$$14.1 \quad H = F^2{}_z c / L^2{}_z$$

Because $F^2{}_z = F^4{}_z = 2P_0$ and $L^2{}_z = L^4{}_z = 4.0$ H take the form given by Equation 14.2 and it is the same throughout the upright tensegrity dome.

$$14.2 \quad H = 2P_0 c / L^2{}_z$$

Forces H are applied to the spatial tensegrity arches by cables A. Because cables A are at angle α, they apply forces V shown in Figure 14.3 to the spatial tensegrity arches.

$$14.3 \quad V = H \tan \alpha = 2P_0 c \tan \alpha / L^2{}_z$$

Because the configuration of the tensegrity arch is based on the suspension cable shown in Figure 14.4, V are "fitted loads" and the tensegrity arch can sustain them in its prestressed configuration.

Forces V apply simultaneously with the prestressing of cables 2 and 4. In the case where the prestressing of these cables is to the level where $F^2{}_z$ and $F^4{}_z$ are equal to $2P_0$, two forces V given by Equation 14.3 apply to bars 8,6 and cable 3 and two forces V apply to cables 5,1 and bar 7. Forces V induce compression to elements 1, 3, 5, 6, 7 and 8 of the tensegrity arch. The vertical component of the compression induced to elements $\Delta F^1{}_z$, $\Delta F^3{}_z$, $\Delta F^5{}_z$, $\Delta F^6{}_z$, $\Delta F^7{}_z$ and $\Delta F^8{}_z$ is given by the force diagrams (a) and (b) shown in Figure 14.6. Since $\tan \alpha = 1.0$, the vertical component of the compression is equal to V.

$$14.4 \quad \Delta F^1{}_z = \Delta F^3{}_z = \Delta F^5{}_z = \Delta F^6{}_z = \Delta F^7{}_z = \Delta F^8{}_z = V = -2P_0 c \tan \alpha / L^2{}_z$$

It is assumed that the magnitude of V is $-\epsilon P_0$ in which ϵ indicates the magnitude of V relative to P_0.

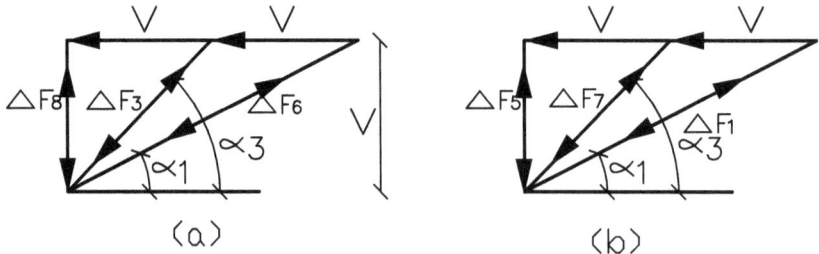

Figure 14.6 Force diagrams considering V.

By using Equation 14.4, $\in$ takes the following form:

$$14.5 \quad \in = 2c \tan \alpha / L^2_z$$

When Equation 14.6 is satisfied, tension is maintained in cables 1, 3 and 5.

$$14.6 \quad P_0 - 2P_0 c \tan \alpha / L^2_z > 0$$

Equation 14.6 implies:

$$14.7 \quad c < L^2_z / (2 \tan \alpha)$$

The compression forces, when induced to cable 1 and bar 6, apply a vertical uplift load of 2V to the top rigid element of the upright tensegrity dome, as shown in Figure 14.7.

To balance these forces in equilibrium and to eliminate moments about a normal axis, a force of $2 \in P_0$ at a distance d given by Equation 14.8 should apply to the top rigid element as shown in Figure 14.7.

$$14.8 \quad d = P_0 c(1 + \in / (2 \in P_0) = c(1 + \in) / (2 \in)$$

Here, d can be used to define f shown in Figure 14.7:

$$14.9 \quad f = 2d - c = c / \in$$

By using Equations 14.5 and 14.9, f takes the following form:

$$14.10 \quad f = L^2_z / (2 \tan \alpha)$$

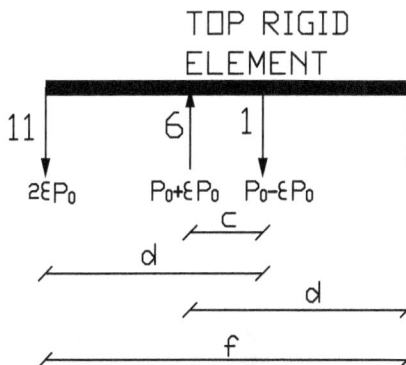

Figure 14.7 Forces applied by cable 1 and bar 6 to the top rigid element.

It is interesting to realize that f is independent of the magnitude of c.

The upright tensegrity dome is designed so that the uplift force of $2V = 2\epsilon P_0$ acting on the top rigid element is carried out by a system of tensegrity rings connected to each other as shown in Figures 14.8 and 14.9.

Members 12, 16, 17, 18 and 19 form the top ring. Members 14, 20, 21, 22 and 23 form the bottom ring. The rings are designed so that points D and E in the middle of cables 12 and 14 coincide with points D and E in the middle of cables 2 and 4 of the tensegrity arches of the upright tensegrity dome. The vertical distance between cable 17 and bar 18 and bar 16 and cable 19 is 2e, as shown in Figure 14.8. The same vertical distance of 2e is between bar 20 cable 23 and cable 21 bar 22 of the lower ring. Cables 11, 13 and 15 connect the two rings to each other. Cables 11, 13 and 15 are parallel to cables 1, 3 and 5, respectively. So cables 11, 12, 13, 14 and 15 form a suspension cable very similar to the one shown in Figure 14.4. Because of the inclination of cables 12 and 14, g_1 and g_2 shown in Figures 14.8 and 14.9 take the values of $g_1 = 3e/2$ and $g_2 = e/2$, as shown in Figures 14.8 and 14.9.

From Figure 14.9, it can be seen that the tensegrity rings are feasible tensegrity rings only where

$$14.11 \quad \tan\alpha > \tan\beta_1$$

Because

$$14.12 \quad \tan\beta_1 = 3e\,/\,f$$

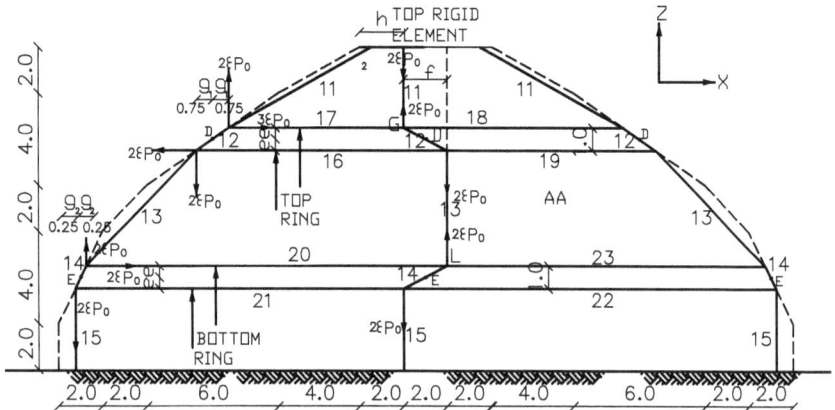

Figure 14.8 The two tensegrity rings sustaining the uplift force UP = $2\epsilon P_0$.

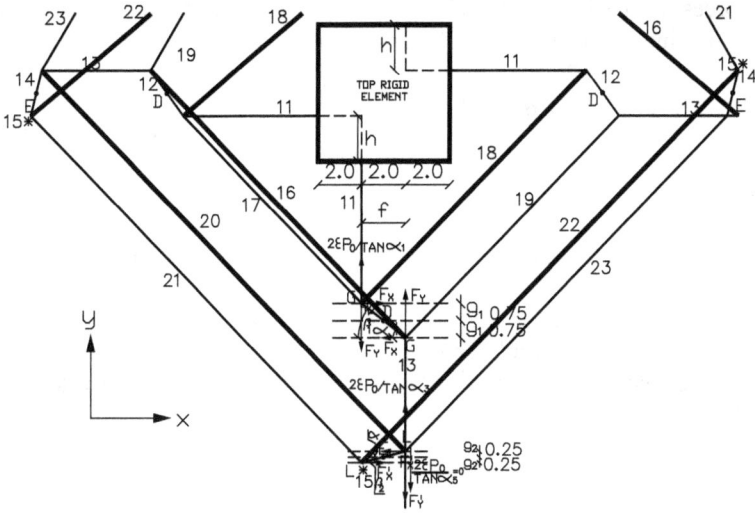

Figure 14.9 Plan of the two tensegrity rings.

Equation 14.11 implies:

$$14.13 \quad e < f \tan \alpha / 3$$

By using Equations 14.10 and 14.13, e takes the value as follows:

$$14.14 \quad e < L_z^2 / 6$$

The bottom ring is a feasible tensegrity ring where

$$14.15 \quad \tan \alpha > \tan \beta_2$$

Because $\tan \beta_2 = e/f$, Equation 14.15 is always satisfied when Equation 14.13 is fulfilled.

Equilibrium of joint G in Figure 14.8 implies:

$$14.16 \quad F^{12}_z = 2 \in P_0$$

By using Equation 14.16, F^{12}_x and F^{12}_y take the following form:

$$14.17 \quad F^{12}_x = f F^{12}_z / (2e) = f \in P_0 / e$$

$$14.18 \quad F^{12}_y = g_1 F^{12}_z / e = 2g_1 \in P_0 / e$$

The forces induced to cables 19 and 17 and to bars 16 and 18 of the top tensegrity ring due to F^{12}_x and F^{12}_y are as follows:

$$F^{16} = - F^{12}_x /(2\cos\alpha) - F^{12}_y /(2\sin\alpha) = - \in P_0 \left(f /(2\cos\alpha) + g_1 /\sin\alpha \right)/e$$

$$F^{19} = F^{12}_x /(2\cos\alpha) - F^{12}_y /(2\sin\alpha) = \in P_0 \left(f /(2\cos\alpha) - g_1 /\sin\alpha \right)/e$$

14.19 $$F^{17} = F^{12}_x /(2\cos\alpha) + F^{12}_y /(2\sin\alpha) = \in P_0 \left(f /(2\cos\alpha) + g_1 /\sin\alpha \right)/e$$

$$F^{18} = - F^{12}_x /(2\cos\alpha) + F^{12}_y /(2\sin\alpha) = - \in P_0 \left(f /(2\cos\alpha) - g_1 /\sin\alpha \right)/e$$

The vertical uplift force of $2\in P_0$ due to cable 11 induces a horizontal force of $2\in P_0/\tan\alpha_1$ to cable 17 and bar 18, as shown in Figure 14.9. This horizontal force induces compression force ΔF given by Equation 14.20 to cable 17 and bar 18.

$$14.20 \quad \Delta F = \in P_0 /(\tan\alpha_1\sin\alpha)$$

It increases the compression in bar 18 and induces compression to cable 17. Tension in cable 17 is maintained when

$$14.21 \quad \in P_0 /(\tan\alpha_1 \sin\alpha) < \in P_0 \left(f /(2\cos\alpha) + g_1 /\sin\alpha \right)/e$$

By using Equations 14.10 and 14.12, Equation 14.21 takes the following form:

$$14.22 \quad e < L^2_z /2$$

It can be seen that where Equation 14.14 is satisfied, Equation 14.22 is fulfilled too. Another critical cable is cable 21. In this case, the horizontal force of $2 \in P_0/\tan\alpha_3$ at joint L induces compression to it. Tension is maintained in cable 21 where

$$14.23 \quad \in P_0 /(\tan\alpha_3 \sin\alpha) < \in P_0 \left(f /(2\cos\alpha) - g_2 /\sin\alpha \right)/e$$

Since $g_2 = e/2$, Equation 14.23 implies:

$$14.24 \quad e < L^2_z /2$$

It can be seen that this condition is similar to the one given by Equation 14.22.

Side view of the upright tensegrity dome with four tensegrity arches and two tensegrity rings is shown in Figure 14.10.

The interaction between the tensegrity arches and the tensegrity rings can be studied by investigating the zone marked by AA in Figure 14.10. Since cables 2 and 12 are straight and intersect at point D, points I, j, k and l are on the same plane. The forces acting in this zone are shown in Figure 14.11.

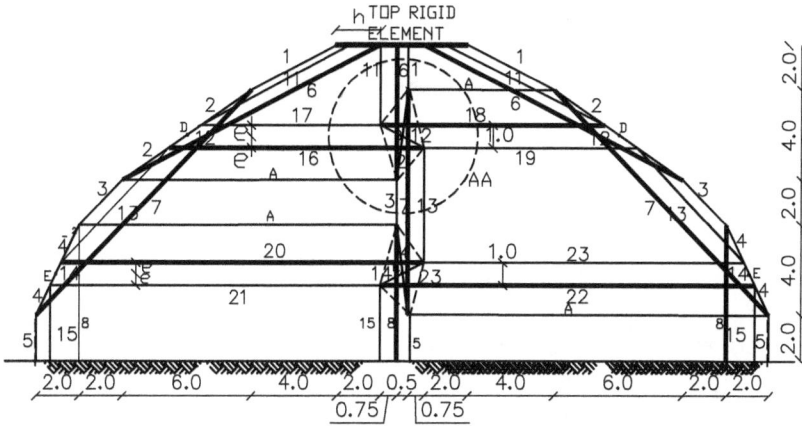

Figure 14.10 Side view of the upright tensegrity dome with four tensegrity arches and two tensegrity rings.

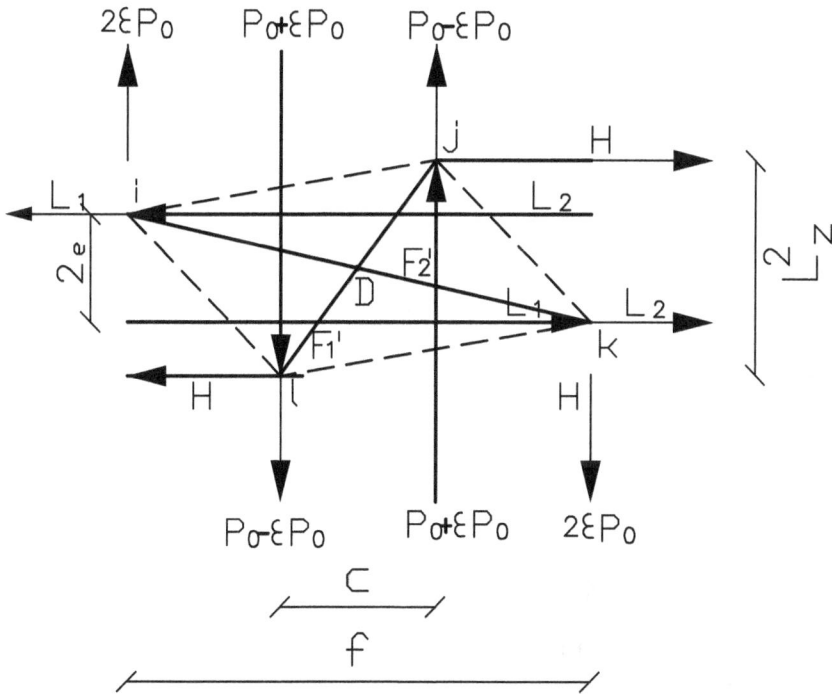

Figure 14.11 Forces at zone AA.

In Figure 14.11, L_1 and L_2 are as follows:

$$14.25 \quad L_1 = F^{17} \cos \alpha; \quad L_2 = F^{18} \cos \alpha$$

F_1' and F_2' are projections on the vertical plane shown in Figure 14.10 of F_2 and F_{12}.

F_2' takes the following form:

$$14.26 \quad \left(F_2'\right)^2 = \left(L_1 + L_2\right)^2 + 4 \, \epsilon^2 \, P_0^{\,2}$$

Equilibrium at node i implies:

$$14.27 \quad \epsilon \, P_0 / \left(L_1 + L_2\right) = e / f$$

By using Equation 14.27, Equation 14.28 takes the following form:

$$14.28 \quad \left(F_2'\right)^2 = \epsilon^2 \, P_0^{\,2} \left(f^2 + 4e^2\right) / e^2$$

F_1' takes the following form:

$$14.29 \quad \left(F_1'\right)^2 = H^2 + 4P_0^{\,2}$$

By using Equation 14.2, Equation 14.29 takes the following form:

$$14.30 \quad \left(F_1'\right)^2 = 4P_0^{\,2} \left(c^2 + \left(L_z^2\right)^2\right) / \left(L_z^2\right)^2$$

The length Lik between joints i and k shown in Figure 14.11 is as follows:

$$14.31 \quad \left(Lik\right)^2 = f^2 + 4e^2$$

The length, Ljl between joints j and l shown in Figure 14.11 is as follows:

$$14.32 \quad \left(Ljl\right)^2 = \left(L_z^2\right)^2 + c^2$$

In the case where

$$14.33 \quad \left(F_1'\right)^2 / \left(F_2'\right)^2 = \left(Lil\right)^2 / \left(Ljk\right)^2$$

The two cables Ljl and Lik can be replaced by the four cables Lij, Ljk, Lkl and Ljl as it is shown in Figure 1.7 in Chapter 1 and by dotted lines in Figure 14.10.

By using Equations 14.28 and 14.30–14.32, Equation 14.33 implies:

$$14.34 \quad \epsilon = 2e / L_z^2$$

By using Equation 14.14, Equation 14.34 implies:

$$14.35 \quad \epsilon < 1/3$$

The dimensions of the various elements of the upright tensegrity dome can be determined according to the following steps:

1. Determining f by using Equation 14.10
2. Assuming e in accordance with Equation 14.14
3. Determining ϵ by using Equation 14.35
4. Using Equation 4.1.9 to determine the magnitude of c

In the case of an upright tensegrity dome with four tensegrity arches shown in Figure 14.10 in which $L_z^2 = 4.0$ and $\alpha = 45°$ f, e, ϵ and c can take the values 2.0, 0.5, 1/4 and 0.5, respectively.

In the case of an upright tensegrity dome with eight tensegrity arches, $\alpha = 22.5°$ and $L_z^2 = L_z^4 = 4.0$ and the other parameters $\alpha_1, \alpha_2, \alpha_3, \alpha_4$ are the same as in the case of the upright tensegrity dome with four arches shown in Figure 14.1. The magnitude of the other parameters of this upright tensegrity dome, f, e, ϵ and c, are determined accordingly and take the values of 4.8, 0.5, 1/4 and 1.2, respectively.

The upright tensegrity dome with eight tensegrity arches is shown in Figures 14.12 and 14.13.

14.2 RECLINING TENSEGRITY DOMES

The method used to design the upright tensegrity dome with eight tensegrity arches presented in the previous chapter can be used to design a reclining tensegrity dome.

The plan of the reclining tensegrity dome is shown in Figure 14.14. Side view is shown in Figure 14.15 and the front view is shown in Figure 14.16.

To ensure that cables 1 and 11 and bar 6 apply no horizontal force to the side gables of the reclining tensegrity dome, one gable should be designed to have no resistance to horizontal forces as shown in Figure 14.15.

In the case of a reclining tensegrity dome, the uplift forces $2 \epsilon P_0$ applied to the top rigid element of the upright tensegrity dome shown in Figure 14.7 are horizontal and apply to the side gables. It is possible to design the side gables to sustain these horizontal forces. In such cases, there is no need of the supporting rings. The dome is an intrinsically tensegrity dome and it takes the shape shown in Figures 14.17–14.19.

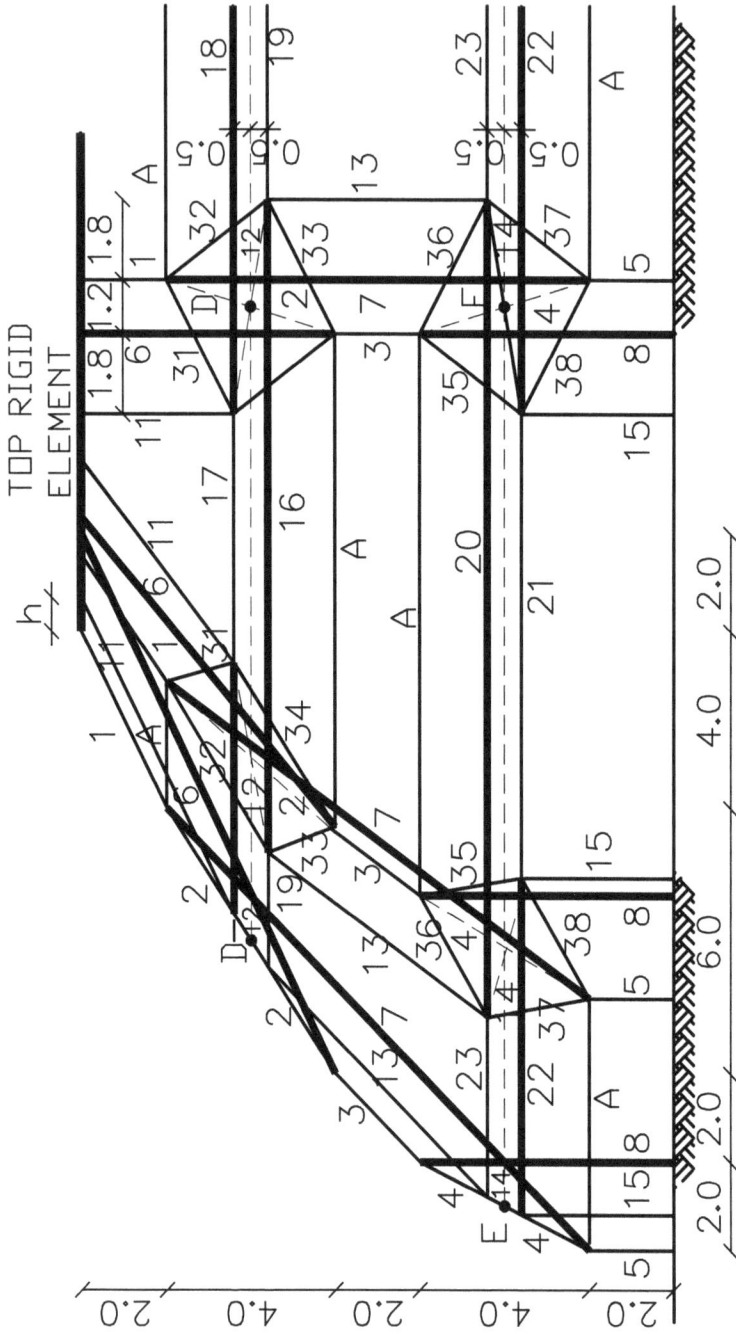

Figure 14.12 Side view of an upright tensegrity dome with eight tensegrity arches.

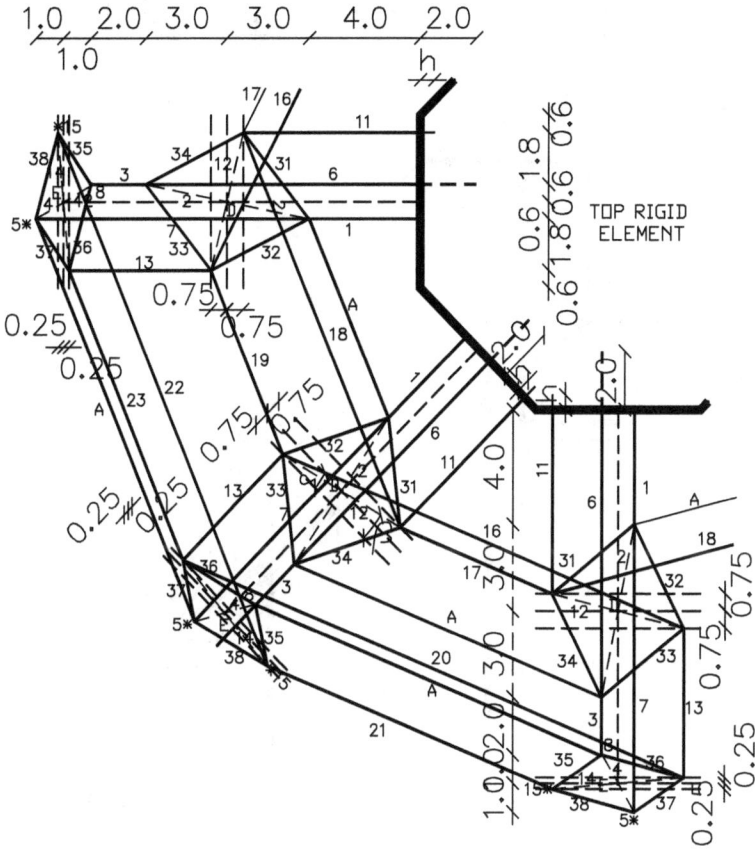

Figure 14.13 Plan of an upright tensegrity dome with eight tensegrity arches.

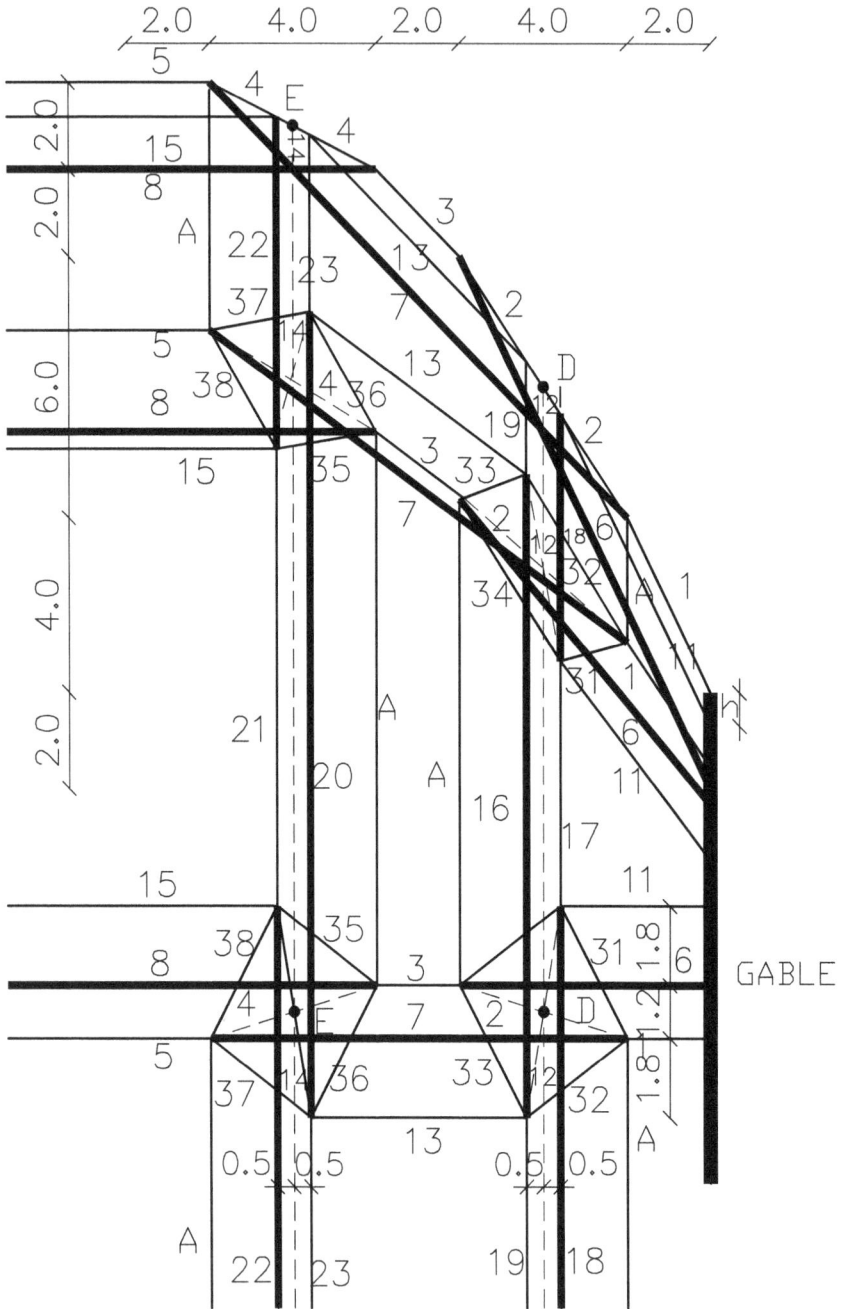

Figure 14.14 Plan of the reclining tensegrity dome.

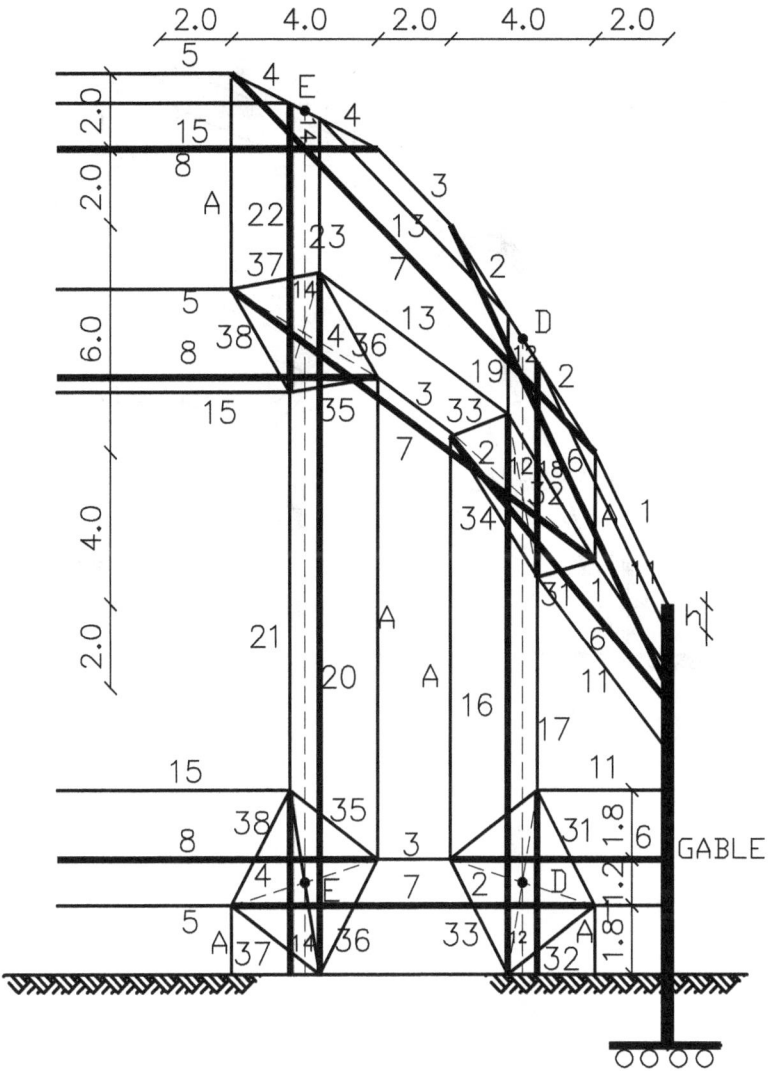

Figure 14.15 Side view of the reclining tensegrity dome.

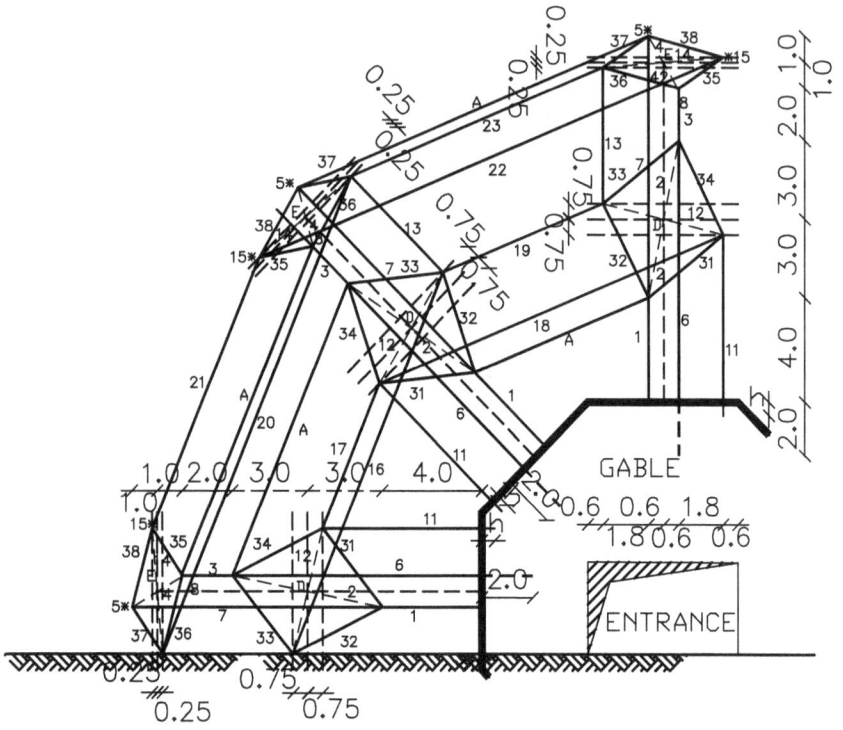

Figure 14.16 Front view of the reclining tensegrity dome.

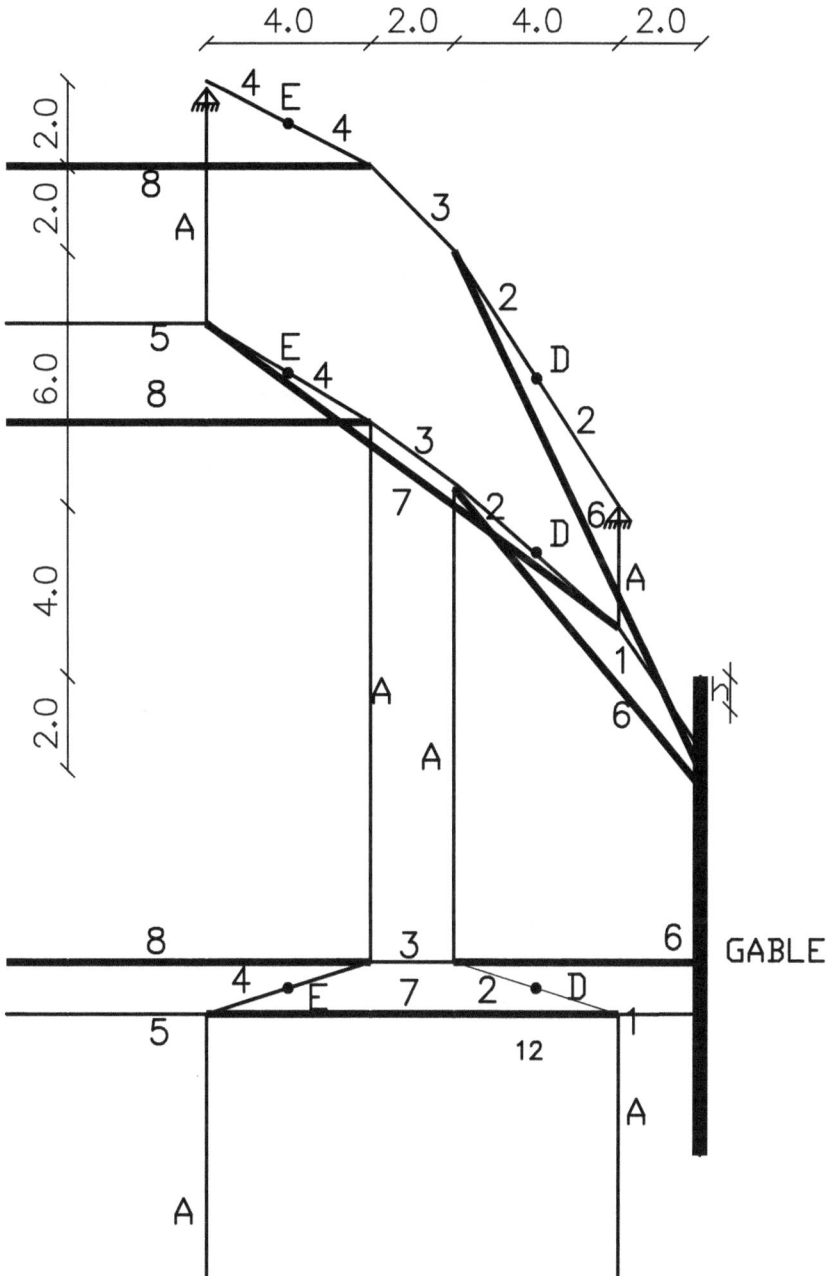

Figure 14.17 Plan of an intrinsically reclining tensegrity dome.

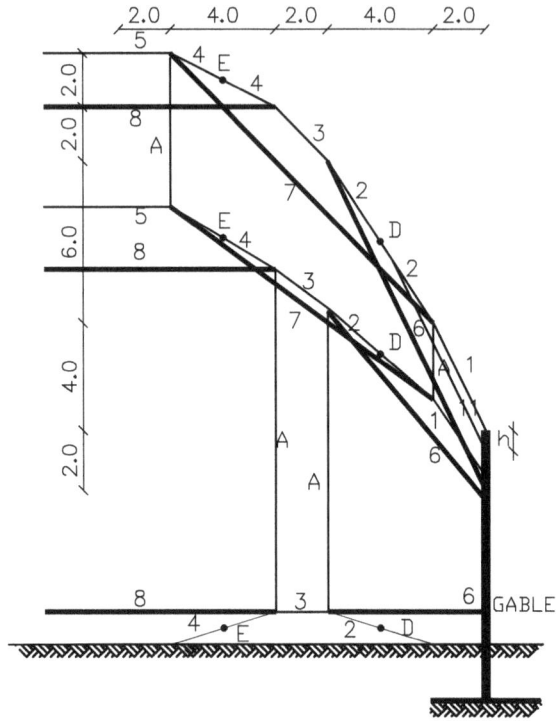

Figure 14.18 Side view of an intrinsically reclining tensegrity dome.

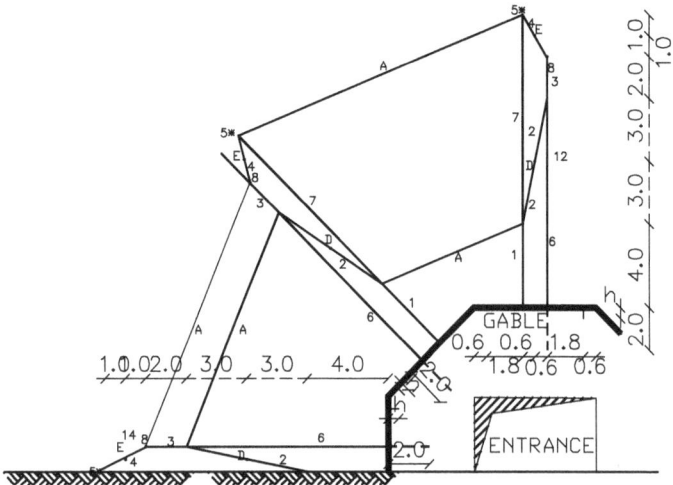

Figure 14.19 Front view of an intrinsically reclining tensegrity dome.

Part 3

Stable tensegrity structures

Tensstable structures

Chapter 15

Two-dimensional tensstable arches

Most of the tensegrity structures are infinitesimal mechanism. These structures can sustain, in the prestressed configuration, a limited type of loads only, the so-called fitted load. When loaded with "fitted load", the nodal displacements of the structure are small similar to common indeterminate structures. To other loads, the tensegrity structure responds as an infinitesimal mechanism. These loads cause relatively large nodal displacements due to rigid body movement of its members until a state of equilibrium is achieved at all nodes. The methods of analysis of the nodal displacement of tensegrity structures are presented in Vilnay (1990).

In the case of dynamical loads, the tensegrity structure responds with small dynamical displacements at high frequencies and large dynamical displacements, that can be observed by the naked eye, at low frequencies.

It is possible, after the prestressing and erecting of an infinitesimal mechanism tensegrity structure, to add bracing cables to the tensegrity structure and to change the tensegrity structure mechanically from an infinitesimal mechanism to a common indeterminate structure with small nodal displacements considering all loading cases, in which the nodal displacements are due to the elasticity of the members of the tensegrity structure only and large nodal displacements due to rigid body movement of members of the tensegrity structure are eliminated.

From structural point of view, changing an infinitesimal mechanism structure to an indeterminate structure is of major importance. In this way, the queer and suspicious tensegrity structure is changed to a structure the structural engineering profession is proficient with.

It is proposed to call this specific family of tensegrity structures tensstable structures (tensegrity + stable) to emphasize the fact that these tensegrity structures respond to loads as common structural engineering structures and not as an infinitesimal mechanisms.

Tensstable structures are mechanically indeterminate, they are composed of a continuous tensile net (cables) and compression elements (bars) which are connected to the tensile net only and not to each other. They are

DOI: 10.1201/9781003370093-18

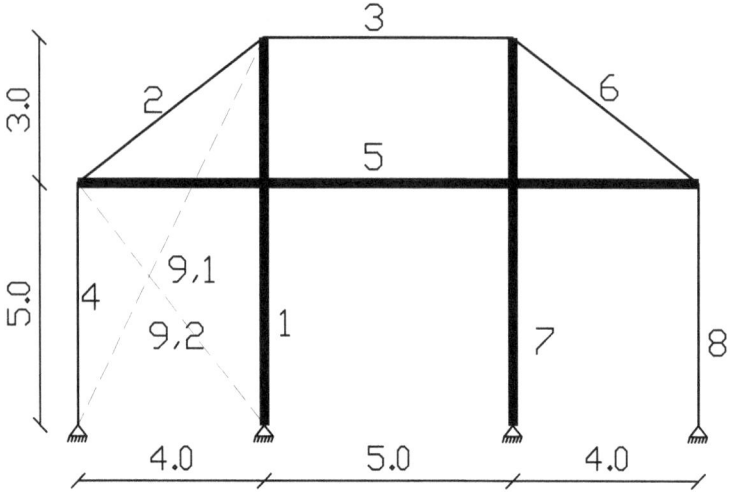

Figure 15.1 Tensegrity arch with additional cables.

constructed with no or only few compression elements that are connected to the foundations. They are prestressable, and prestressing induces tension to the cables and compression to the bars.

For example, in the case of the tensegrity arch shown in Figure 4.2, two additional cables, 9,1 and 9,2 shown in Figure 15.1, are added to the tensegrity arch after it was prestressed and erected. These two bracing cables change the tensegrity arch to a tensstable structure.

Matrix A of the equilibrium equations of forces at the nodes of the tensstable arch shown in Figure 15.1 is similar to the one given by Equation 4.1. Since the bracing is by cables, and a cable is active only when under tension, matrix A of the tensstable arch takes the form given by Equation 15.1 in the case where the external load induces tension to cable 9,1:

$$
15.1 \quad A = \begin{vmatrix}
-1 & -3.5 & 0 & 0 & 0 & 0 & 0 & 0 & -8/\sqrt{80} \\
0 & -4/.5 & 1 & 0 & 0 & 0 & 0 & 0 & -4/\sqrt{80} \\
0 & 3/5 & 0 & -1 & 0 & 0 & 0 & 0 & 0 \\
0 & 4/5 & 0 & 0 & 1 & 0 & 0 & 0 & 0 \\
0 & 0 & 0 & 0 & 0 & -3/5 & -1 & 0 & 0 \\
0 & 0 & -1 & 0 & 0 & 4/5 & 0 & 0 & 0 \\
0 & 0 & 0 & 0 & 0 & 3/5 & 0 & -1 & 0 \\
0 & 0 & 0 & 0 & -1 & -4/5 & 0 & 0 & 0
\end{vmatrix}
\begin{matrix}
L1 \\ L2 \\ L3 \\ L4 \\ L5 \\ L6 \\ L7 \\ L8
\end{matrix}
$$

In the case where the external load induces tension to cable 9,2, matrix A takes the form given by Equation 15.2:

$$
15.2 \quad A = \begin{array}{|ccccccccc|l}
-1 & -3.5 & 0 & 0 & 0 & 0 & 0 & 0 & 0 & L1 \\
0 & -4/.5 & 1 & 0 & 0 & 0 & 0 & 0 & 0 & L2 \\
0 & 3/5 & 0 & -1 & 0 & 0 & 0 & 0 & -5/\sqrt{41} & L3 \\
0 & 4/5 & 0 & 0 & 1 & 0 & 0 & 0 & 4/\sqrt{41} & L4 \\
0 & 0 & 0 & 0 & 0 & -3/5 & -1 & 0 & 0 & L5 \\
0 & 0 & -1 & 0 & 0 & 4/5 & 0 & 0 & 0 & L6 \\
0 & 0 & 0 & 0 & 0 & 3/5 & 0 & -1 & 0 & L7 \\
0 & 0 & 0 & 0 & -1 & -4/5 & 0 & 0 & 0 & L8
\end{array}
$$

It can be seen that in both matrices A given by Equations 15.1 and 15.2, L2 + L4 + L6 + L8 ≠ 0, different from matrix A given by Equation 4.1 where Equation 4.2 is satisfied. Adding cable 9,1 or 9,2 to the tensegrity arch in Figure 4.2 changed the rank of matrix A to be equal to the number of equilibrium equations. Because there are more unknown forces than equilibrium equations, the tensstable arch shown in Figure 15.1 is an indeterminate structure. Adding cables 9,1 and 9,2 change the tensegrity arch shown in Figure 4.2 from an infinitesimal mechanism to a tensstable structure.

In each loading case, methods of analysis of indeterminate structures can be used to determine the forces in the tensstable structure members. The analysis should be carried out once by using matrix A given by Equation 15.1 and once by using matrix A given by Equation 15.2. The valid analysis is the one associated with matrix A which results with tension in the relevant bracing cable.

Not always adding bracing cables to the tensegrity arch changes it into a tensstable structure. For example, the case where the bracing cables are added to the tensegrity arch shown in Figure 4.2, as shown in Figure 15.2.

In the case where only cable 9,3 is active, matrix A takes the following form:

$$
15.3 \quad A = \begin{array}{|ccccccccc|l}
-1 & -3.5 & 0 & 0 & 0 & 0 & 0 & 0 & 0 & L1 \\
0 & -4/.5 & 1 & 0 & 0 & 0 & 0 & 0 & 0 & L2 \\
0 & 3/5 & 0 & -1 & 0 & 0 & 0 & 0 & 3/\sqrt{90} & L3 \\
0 & 4/5 & 0 & 0 & 1 & 0 & 0 & 0 & 9/\sqrt{90} & L4 \\
0 & 0 & 0 & 0 & 0 & -3/5 & -1 & 0 & -3/\sqrt{90} & L5 \\
0 & 0 & -1 & 0 & 0 & 4/5 & 0 & 0 & -9/\sqrt{90} & L6 \\
0 & 0 & 0 & 0 & 0 & 3/5 & 0 & -1 & 0 & L7 \\
0 & 0 & 0 & 0 & -1 & -4/5 & 0 & 0 & 0 & L8
\end{array}
$$

It can be seen that in this tensegrity arch, the following relationship between the lines of matrix A, L2 + L4 + L6 + L8 = 0, is still valid. This fact indicates that by adding cable 9,3, the tensegrity arch still remains a tensegrity arch and was not changed to a tensstable arch. The same is true when cable 9,4 is added to the tensegrity arch.

To ensure the transformation of the tensegrity arch to a tensstable arch, it is preferable to add bracing at every possible location as shown in Figure 15.3.

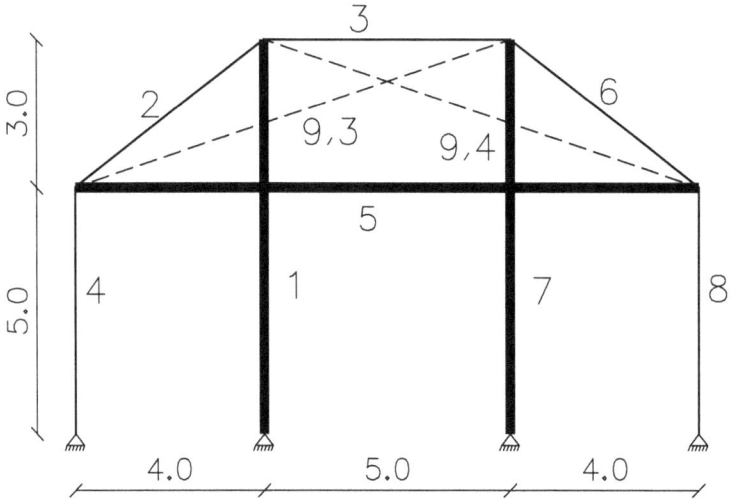

Figure 15.2 Tensegrity arch with bracing cables.

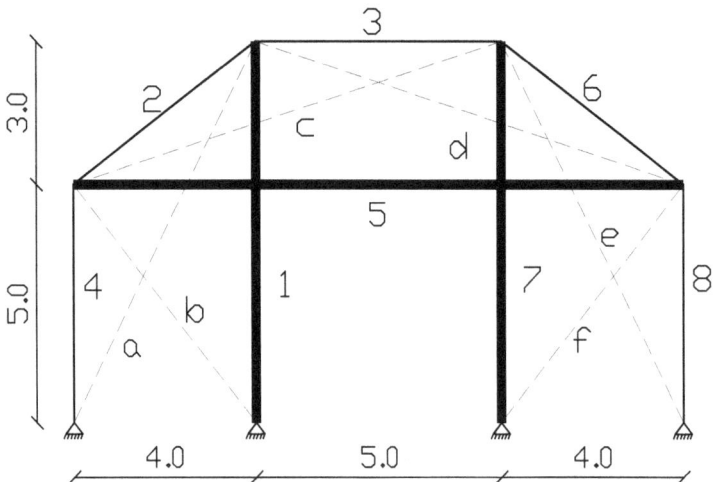

Figure 15.3 Bracing tensstable arch.

Bracing can be added after the prestressing and erection of the tensegrity structure. The bracing should be tight to ensure prompt response to the loads applied to the tensstable structure. In this case, when the tensstable structure is loaded, the load induces tension to some of the bracing cables and some will be slack and not tight.

It is also possible to prestress the bracing. In this case when properly prestressed, external load reduces the tension in some bracing cables and increases the tension in others. The prestressing is designed to the level required to keep all cables tight and straight under tension.

To consider the effect of prestressing the bracing cables, a typical bracing which takes the form shown in Figure 15.4 is studied.

Elements a-d and c-b in Figure 15.4 are the bracing cables and elements a-b, a-c, c-d and b-d are members of the tensegrity structure. In the case where elements a-c and b-d are parallel to each other, a-b-c-d is a trapezium. The case of a tensegrity structure in the shape of a trapezium was discussed

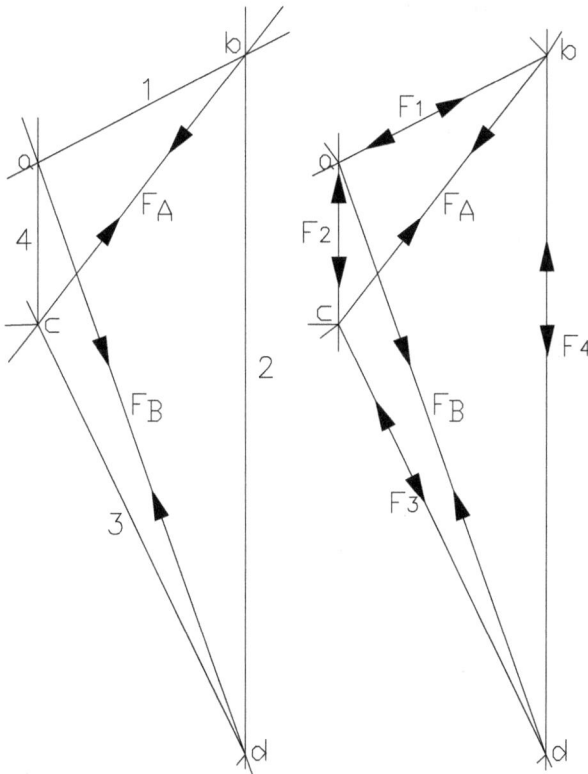

Figure 15.4 Typical bracing.

in Chapter 1 and its force diagram is shown in Figure 1.8. It can be seen that the bracing forces can be in equilibrium with forces in the members of the already prestressed tensegrity structure in the following three cases: prestressing cable A, prestressing cable B and by prestressing simultaneously cables A and B by F_A and F_B where Equation 15.4 is satisfied.

$$15.4 \quad F_A / F_B = L_A / L_B$$

Here L_A and L_B are the lengths of the bracing cables, cable A and B, respectively. The "mechanical method" presented in Chapter 1 implies that in these three cases, the bracing cables apply "fitted load" to the tensegrity structure.

After bracing cables were added at all possible locations, few methods can be adopted to prestress the bracing cables.

When the so-called piecewise method is adopted, every bracing is installed and prestressed in its turn. For example, in the case of the tensegrity arch shown in Figure 15.3, it is possible to install and prestress first: bracing of cables c and d. In the case where cable c is prestressed, matrix A considering cable d takes the following form:

$$15.5 \quad A = \begin{vmatrix} -1 & -3.5 & 0 & 0 & 0 & 0 & 0 & 0 & -3/\sqrt{90} \\ 0 & -4/.5 & 1 & 0 & 0 & 0 & 0 & 0 & 9/\sqrt{90} \\ 0 & 3/5 & 0 & -1 & 0 & 0 & 0 & 0 & 0 \\ 0 & 4/5 & 0 & 0 & 1 & 0 & 0 & 0 & 0 \\ 0 & 0 & 0 & 0 & 0 & -3/5 & -1 & 0 & 0 \\ 0 & 0 & -1 & 0 & 0 & 4/5 & 0 & 0 & 0 \\ 0 & 0 & 0 & 0 & 0 & 3/5 & 0 & -1 & 3/\sqrt{90} \\ 0 & 0 & 0 & 0 & -1 & -4/5 & 0 & 0 & -9/\sqrt{90} \end{vmatrix}$$

The load Q applies to the tensegrity arch by prestressing cable d to the level of P^C_0 and takes the following form:

$$15.6 \quad Q = P^C_0 / \sqrt{90} \begin{vmatrix} 0 \\ 0 \\ -3 \\ -9 \\ 9 \\ 3 \\ 0 \\ 0 \end{vmatrix}$$

It can be seen that this load is "fitted load" since the condition of L2 + L4 + L6 + L8 = 0 is satisfied and so the tensegrity arch can sustain it in its given configuration. But to determine the change of the forces in the tensegrity, arch member required an extensive tensegrity structures analysis.

Matrix A of the tensegrity arch with the prestressed cables c and d takes the following form:

$$
15.7 \quad \mathbf{A} = \begin{vmatrix}
-1 & -3.5 & 0 & 0 & 0 & 0 & 0 & 0 & 0 & -3/\sqrt{90} \\
0 & -4/5 & 1 & 0 & 0 & 0 & 0 & 0 & 0 & 9/\sqrt{90} \\
0 & 3/5 & 0 & -1 & 0 & 0 & 0 & 0 & 3/\sqrt{90} & 0 \\
0 & 4/5 & 0 & 0 & 1 & 0 & 0 & 0 & 9/\sqrt{90} & 0 \\
0 & 0 & 0 & 0 & 0 & -3/5 & -1 & 0 & -3/\sqrt{90} & 0 \\
0 & 0 & -1 & 0 & 0 & 4/5 & 0 & 0 & -9/\sqrt{90} & 0 \\
0 & 0 & 0 & 0 & 0 & 3/5 & 0 & -1 & 0 & 3/\sqrt{90} \\
0 & 0 & 0 & 0 & -1 & -4/5 & 0 & 0 & 0 & -9/\sqrt{90}
\end{vmatrix}
$$

It can be seen that also in this case L2 + L4 + L5 + L8 = 0; this implies that even after adding cables c and d, the tensegrity arch is still a tensegrity structure and the effect of load applied to the tensegrity structure should follow tensegrity structure analysis.

In the second stage, bracing of cables a and b shown in Figure 15.3 is added. In the case where this bracing is prestressed by force applied by cable b, matrix A considering cables a, c, and a takes the following form:

15.8

$$
\mathbf{A} = \begin{vmatrix}
-1 & -3/5 & 0 & 0 & 0 & 0 & 0 & 0 & 0 & -3/\sqrt{90} & -8/\sqrt{90} \\
0 & -4/5 & 1 & 0 & 0 & 0 & 0 & 0 & 0 & 9/\sqrt{90} & -4/\sqrt{90} \\
0 & 3/5 & 0 & -1 & 0 & 0 & 0 & 0 & 3/\sqrt{90} & 0 & 0 \\
0 & 4/5 & 0 & 0 & 1 & 0 & 0 & 0 & 9/\sqrt{90} & 0 & 0 \\
0 & 0 & 0 & 0 & 0 & -3/5 & -1 & 0 & -3/\sqrt{90} & 0 & 0 \\
0 & 0 & -1 & 0 & 0 & 4/5 & 0 & 0 & -9/\sqrt{90} & 0 & 0 \\
0 & 0 & 0 & 0 & 0 & 3/5 & 0 & -1 & 0 & 3/\sqrt{90} & 0 \\
0 & 0 & 0 & 0 & -1 & -4/5 & 0 & 0 & 0 & -9/\sqrt{90} & 0
\end{vmatrix}
$$

It can be seen that condition L2 + L4 + L6 + L8 = 0 is not satisfied: the tensegrity structure was changed to an indeterminate one. The load P^B_0 applies by cable b to this indeterminate tensstable arch.

$$15.9 \quad Q = P^B{}_0 / \sqrt{41} \begin{vmatrix} 0 \\ 0 \\ -5 \\ 4 \\ 0 \\ 0 \\ 0 \\ 0 \end{vmatrix}$$

The analysis of the forces induced to the tensstable arch can follow the well-known methods of analysis of indeterminate structures. Also, cables e and f can be added and, for example, cable f can be prestressed accordingly, the effect of prestressing cable f can be found by using indeterminate methods of analysis. The disadvantage of the "piecewise method" is the extensive use of tensegrity structure methods of analysis.

The second method that can be used to change the tensegrity structure to a tensstable structure is the so-called indeterminate method. At the first stage of the "indeterminate method", the tensegrity structure is prestressed and erected and all the designed bracing cables are installed. At this stage, one bracing can be prestressed by prestressing one of its cables. The forces induced by the prestressing to the other bracing can be determined by using methods of analysis of indeterminate structures and there is no need to use tensegrity structure analysis. In the case in which indeterminate methods of analysis fail, it is an indication that the bracing added are short of the level required to change the tensegrity structure to a tensstable one. Bracing cables which are under compression are omitted and the analysis is resumed again and the tension in the proper bracing cables can be established. Then all bracing that were omitted are restored and the procedure is resumed. At the end of this stage, only cables in bracing that were not prestressed already are omitted. Prestressing at each stage is to the level required to maintain tension in all cables. The procedure continues until all bracing cables are under tension. Care should be taken to ensure that tension is maintained in all cables of the original tensegrity structure. The advantage of the "indeterminate method" is that only indeterminate structure analysis is used.

In the case of the design of the tensstable arch shown in Figure 15.3, the "indeterminate method" can follow these stages. At the first stage, it is assumed that all cables a, b, c, d, e and f are active and cable d is prestressed. In the case where cables b and f are under compression, analysis is resumed considering cables a, c and e only. After the forces in these cables are established, the second step of the analysis can be resumed. Cables f and b are reinstalled and cable b is prestressed considering all bracings a–f; if tension is induced into cable f, prestressing the bracing is successful.

In the case where prestressing simultaneously of many cables is not an object, the so-called "instant method" can be adopted. When this method is followed, the tensegrity structure is prestressed simultaneously with the bracing cables. Equation 15.4 is observed to determine the magnitude of the forces apply to every one of the bracing cables. The designer is free to determine the magnitude of the prestressing forces of the tensegrity structure and each bracing. Since the bracings apply "fitted load" to the tensegrity arch, the forces induced into the tensstable structure are determined by using simple equilibrium only.

In the case of the tensegrity arch shown in Figure 15.3, a possible assumption of the forces in cables a, b, c, d, e and f is $F_a = \sqrt{80}\ P^1_0$, $F_b = \sqrt{41}\ P^1_0$, $F_c = F_d = \sqrt{10}\ P^1_0$, $F_e = \sqrt{20}\ P^1_0$, $F_f = \sqrt{10.25}\ P^1_0$ and it is assumed that the tensegrity arch is simultaneously prestressed by inducing $8P^1_0$ to cable 4. The prestressing forces induced to all members of the tensstable arch can be found by using equilibrium only and are shown in the force diagram in Figure 15.5.

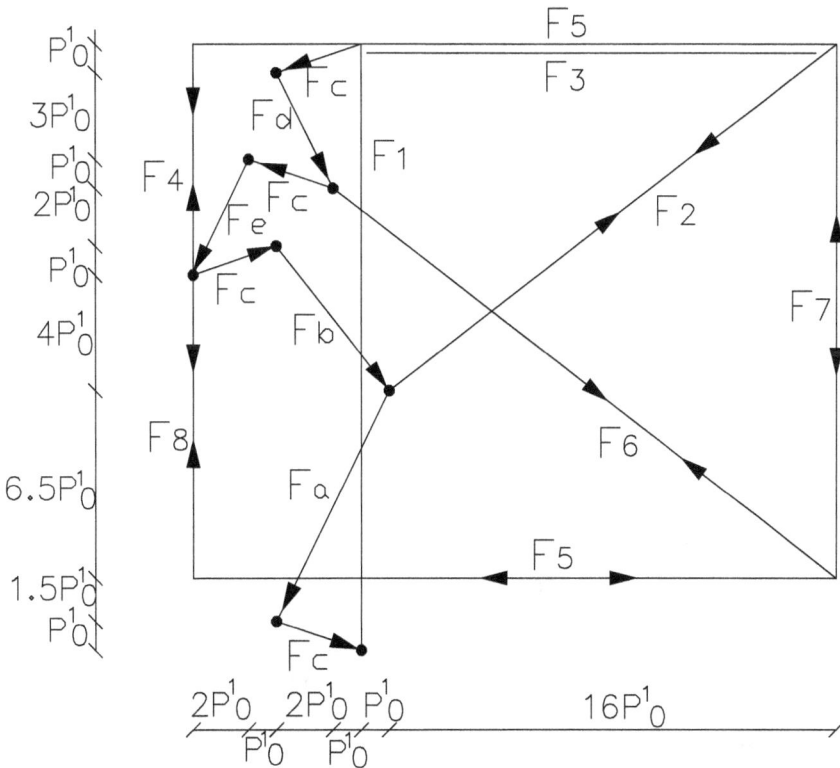

Figure 15.5 Forces induced into the tensstable arch by using the instant method.

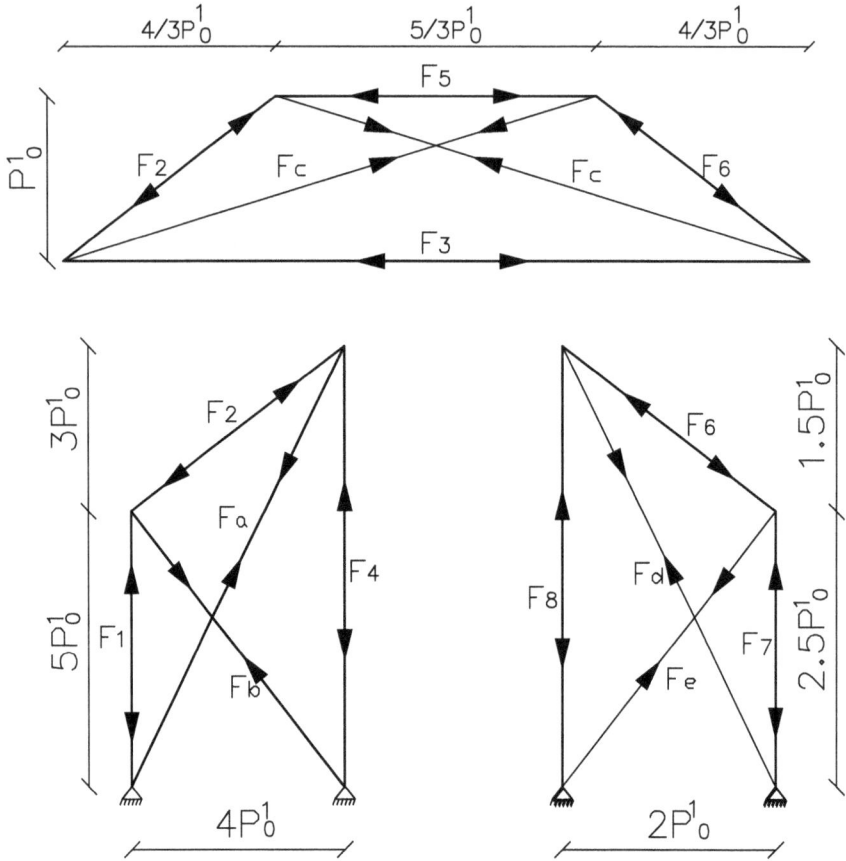

Figure 15.6 Forces induced by prestressed bracings.

The forces in the force diagram shown in Figure 15.5 can be seen as super-position of the forces of the force diagram of the initial prestressing and the forces of the force diagrams of the forces induced by prestressing the bracing cables of the tensegrity arch. The initial prestressing force diagram is shown in Figure 4.3. The fact that the prestressing forces of the bracing cables satisfy Equation 15.4 indicates that the forces in each bracing can be in equilibrium with the relevant members of the tensegrity arch, as shown in Figure 15.6.

By using Figure 15.6, it can be seen that prestressing of the bracing induces compression of $8P^1_0$ to cable 4. In order to maintain tension of $8P^1_0$ in cable 4 during the simultaneous prestressing as required, the prestressing force of $6P_0$ shown in Figure 4.3 should satisfy Equation 15.10.

$$15.10 \quad 6P_0 - 8P^1_0 = 8P^1_0 \quad P_0 = 8P^1_0 / 3$$

It is easy to realize that the forces of the force diagram in Figure 15.5 is a superposition of the forces in the force diagram of the initial prestressing and the forces in the force diagrams of the forces induced by prestressing the bracing cables of the tensegrity arch. For example, it can be seen that the magnitude of the prestressing force in bar 5 in Figure 15.5 is $23P^1_0$ which is equal to the compression force induced by the initial prestressing of $64P^1_0/3$ shown in Figure 4.3 plus the compression of $5P^1_0/3$ induced by the bracing shown in Figure 15.6.

In assuming the required prestressing forces to erect a tensstable structure, the designer is free to impose relevant conditions to the magnitude of the prestressing forces of the bracing and can use, for example, optimization methods if found necessary.

Chapter 16

Three-dimensional tensstable structures

As in the case of two-dimensional tensegrity structure presented in Chapter 15, three-dimensional tensegrity structures can also be changed to tensstable structures by adding proper bracing.

In most cases, possible location of the bracing cables is obvious. For example, in the case of the level tensegrity vault presented in Figures 8.1–8.3, the bracing can be placed as shown by the dotted lines in Figure 16.1.

In the case where the bracing is prestressed following the "instant method" described in Chapter 15, the prestressing of the bracing of the level tensstable vault is according to Equation 15.4.

Proposed bracing to change the vertical tensegrity cylindrical shell shown in Figures 9.1 and 9.2 to a vertical tensstable cylindrical shell is shown in Figure 16.2.

Also, in this case, the prestressing of the bracing cables should follow Equation 15.4 when appropriate.

Proposed bracing to change the upright tensegrity dome shown in Figure 14.10 to an upright tensstable dome is shown in Figure 16.3.

Also in this case the bracing cables should satisfy Equation 15.4 when appropriate.

An interesting case is the slim vertical tensegrity dome shown in Figures 12.1 and 12.2. This slim vertical tensegrity dome presented in Chapter 11 can be changed to a slim vertical tensstable dome by adding the bracing shown in Figure 16.4.

Also, in the case of the slim vertical tensstable dome, prestressing the bracing cables should follow Equation 15.4 when appropriate.

In changing the slim vertical dome to a tensstable slim vertical dome, it is possible to eliminate the diagonal cables associated with the uplift forces applied to the top rigid element shown in Figure 12.6. To determine the effect of eliminating them, it is possible after installing the bracing to apply the uplift forces to the slim vertical tensstable dome. The forces induced to the bracing indicate the magnitude of prestressing required to keep all cables of the slim vertical tensstable dome in tension.

DOI: 10.1201/9781003370093-19

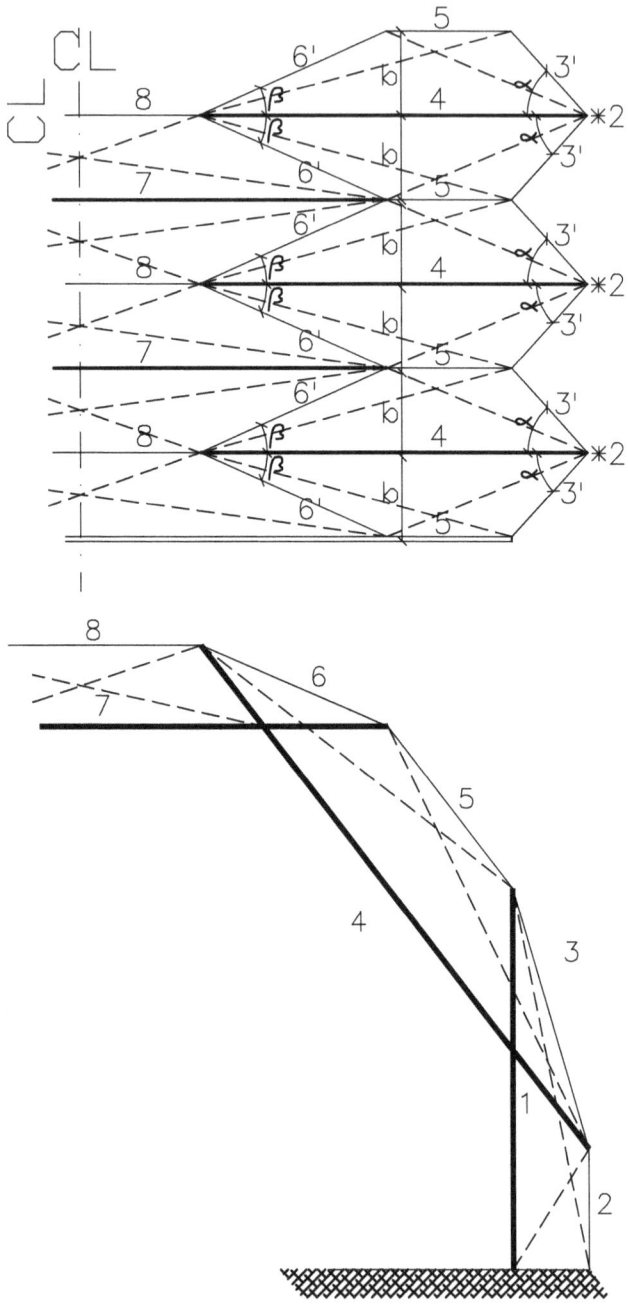

Figure 16.1 Level tensstable vault.

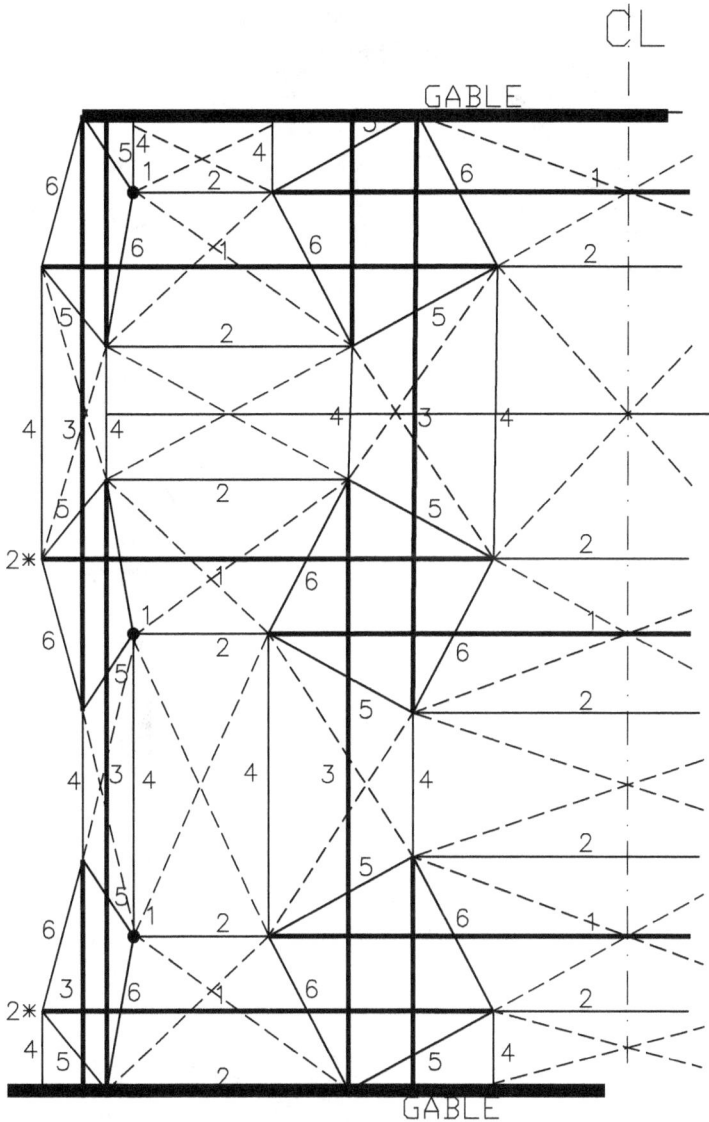

Figure 16.2 Vertical tensstable cylindrical shell.

Because tensstable structures are mechanically indeterminate structures, structural engineers who are used to and familiar with may adopt them without hesitation. It is hoped that imaginative structural engineers will find ingenious ways to follow the methods of design described in this book to develop tensstable structure for mundane use.

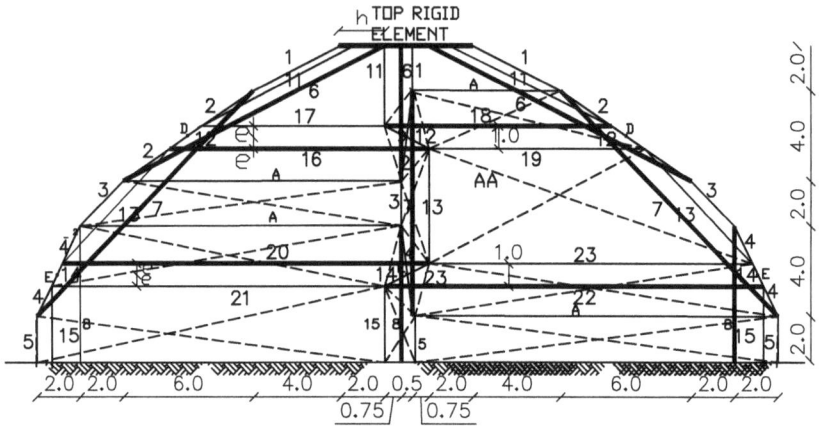

Figure 16.3 Upright tensstable dome.

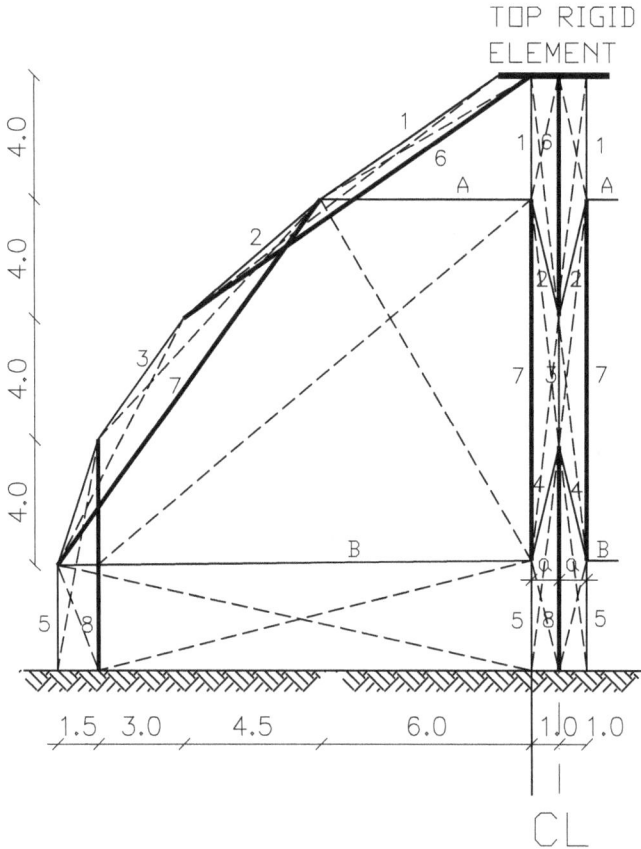

Figure 16.4 Slim vertical tensstable dome.

References

Aldrich, J.B., Skelton, R.E., Kreutz-Delgado, K. (2003) Control synthesis for a class of light and agile robotic tensegrity structures. In: Proceedings of the IEEE American Control Conference, Denver, CO.

Applewhite, E.J. (1986) Synergetics dictionary: The mind of Buckminster Fuller: With an introduction and appendices. New York: Garland.

ArchDaily (2014a) First-place winner of Santiago Landmark Competition: Smiljan Radic + Gabriela Medrano + Ricardo Serpell. Available at: www.archdaily.com/478897/first-place-winner-of-santiago-landmark-competition-smiljan-radic-gabriela-medrano-ricardo-serpell

ArchDaily (2014b) Students of ball state construct parametric tensegrity structure for local art fair. Available at: www.archdaily.com/553311/students-of-ball-state-construct-parametric-tensegrity-structure-for-local-art-fair

ARUP (2022) World's first tensegrity bridge. Available at: www.arup.com/projects/kurilpa-bridge

AurelVR (2008) Dubai tensegrity tower. Available at: www.aurelvr.com/content/dubai-tensegrity-tower

Barnes, M.R. (1999) Form finding and analysis of tension structures by dynamic relaxation. International Journal of Space Structures, 14(2), 89–104.

Bauer, J., Kraus, J.A., Crook, C., Rimoli, J.J., Valdevit, L. (2021) Tensegrity metamaterials: Toward failure-resistant engineering systems through delocalized deformation. Advanced Materials, 33(10), 2005647.

BIRDAIR (2022) Georgia Dome. Available at: www.birdair.com/birdair-portfolio/georgia-dome/

Bouderbala, M., Motro, R. (1998) Folding tensegrity systems. In: Proceedings of IUTAM/IASS Symposium on Deployable Structures: Theory and Applications, Cambridge.

Bruce, J., Caluwaerts, K., Iscen, A., Sabelhaus, A.P., SunSpiral, V. (2014) Design and evolution of a modular tensegrity robot platform. In: 2014 IEEE International Conference on Robotics and Automation (ICRA), pp. 3483–3489. Hong Kong: IEEE. Available at: https://doi.org/10.1109/ICRA.2014.6907361

Burkhardt, R.W. Jr. (2008) A practical guide to tensegrity design. Available at: https://www.angelfire.com/ma4/bob_wb/tenseg.pdf

Burkhardt, R.W. Jr. (2011) Synergetics: Definition and classification of tensegrities, 20 October. Available at: https://bobwb.tripod.com/synergetics/tensegrity/index.html

Calladine, C.R. (1978) Buckminster Fuller's tensegrity structures and Clerk Maxwell's rules for the construction of stiff frames. International Journal of Solids and Structures, 14, 161–172.

Calladine, C.R., Pellegrino, S. (1991) First order infinitesimal mechanisms. International Journal of Solids and Structures, 27(4), 505–515.

Calladine, C.R., Pellegrino, S. (1992) Further remarks on first order infinitesimal mechanisms. International Journal Solids and Structures, 29, 2119–2122.

Caspar, D.L.D., Klug, A. (1962) Physical principles in the construction of regular viruses. In: Proceedings of Cold Spring Harbor symposium on quantitative biology (Vol. 27), pp. 1–24. Cold Spring Harbor Laboratory Press. doi:10.1101/sqb.1962.027.001.005

Chassagnoux, A. (2006) David Georges Emmerich professor of morphology. International Journal of Space Structures, 21(1), 59–71.

Connelly, R. (1980) The rigidity of certain cabled networks and the second order rigidity of arbitrarily triangulated convex surfaces. Advances in Mathematics, 37, 272–299.

Connelly, R. (1982) Rigidity and energy. Inventiones Mathematicae, 66, 11–33.

Connelly, R., Back, A. (1998) Mathematics and tensegrity. American Scientist, 86(2), 453–491.

Connelly, R., Whiteley, W. (1996) Second order rigidity and prestress stability for tensegrity frameworks. Discrete Mathematics, 9(3), 453–491.

Coplans, J. (1967) An interview with Kenneth Snelson. *Artforum*, March, pp. 46–49.

Defossez, M. (2003) Shape memory effect in tensegrity structures. Mechanics Research Communications, 30(4), 311–316.

Dos Santos, F.A., Rodrigues, A., Micheletti, A. (2015) Design and experimental testing of an adaptive shape-morphing tensegrity structure, with frequency self-tuning capabilities, using shape-memory alloys. Smart Materials and Structures, 24(10), 105008.

Edmondson, A.C. (1986) Geodesic reports: The deresonated tensegrity dome. Synergetica: Journal of Synergetics, 1(4).

Emmerich, D.G. (1959) "Carpentes Perles", (or "Pearl Frameworks"). Institut National de la Propriete Indusriéllé (Registration No. 59423), Institut National de la Propriété Industrielle, Courbevoie, France, 29 May.

Emmerich, D.G. (1963) Constructions de Reseaux Autotendants. Brevet N°. 1.377.290. Ministére de L'Industrie, Paris, République Française, 10 Avril.

Emmerich, D.G. (1967a) Cours de Géométrie Constructive Morphologie École Nationale Supérieure des Beaux Arts. Paris: Centre de diffusion de La Grande Masse.

Emmerich, D.G. (1967b) "Réseaux", in space structures: A study o methods and developments in three-dimensional construction. In: Proceedings of the International Conference on Space Structures, University of Surrey, Guildford, 1966, pp. 1059–1072. Guildford: Blackwell Scientific Publications.

Emmerich, D.G. (1988) Structures tendues et autotendantes, Monographies de Geometrie Constructive. Paris: Editions de l'Ecole d'Architecture de Paris La Villette.

Estrada, G.G., Bungartz, H.J., Mohrdiceck, C. (2006) Numerical form-finding of tensegrity structures. International Journal of Solids and Structures, 43(22–23), 6855–6868.

Feng, Y. (2019) Design of freeform membrane-tensegrity structure. MSc Thesis, Department of Architecture, Aalto University School of Arts, Design and Architecture.

Available at: https://aaltodoc.aalto.fi/bitstream/handle/123456789/41455/master_Feng_Ye_2019.pdf?isAllowed=y&sequence=1

Fuller, R. (1955) Buckminster. Architecture out of the laboratory. Dimension (Vols. 1–4). Ann Arbor, MI: University of Michigan, College of Architecture and Design.

Fuller, R. (1961) Buckminster. "Tensegrity", portfolio and art news annual, (4), 112–127, 144, 148. Available at: www.rwgrayprojects.com/rbfnotes/fpapers/tensegrity/tenseg01.html

Fuller, R. (1962) Buckminster. Tensile-integrity structures, United States Patent No. 3063521.

Fuller, R.B. (1975) Synergetics: Exploration in the geometry of thinking. New York: Collier Macmillan Publishers. Available at: www.rwgrayprojects.com/synergetics/synergetics.html

Fuller, R.B., Marks, R.W. (1973) The dymaxion world of Buckminster Fuller. Garden City, NY: Anchor Books.

Furuya, H. (1992) Concept of deployable tensegrity structures in space applications. International Journal of Space Structures, 7(2), 143–151.

Gan, B.S. (2020) Tensegrity in biological application: Cellular tensegrity. In: Computational modeling of tensegrity structures: Art, nature, mechanical and biological systems. Cham: Springer International Publishing, pp. 193–208.

Ganga, P.L., Micheletti, A., Podio-Guidugli, P., Scolamiero, L., Tibert, A.G., Zolesi, V. (2016) Tensegrity rings for deployable space antennas: Concept, design, analysis, and prototype testing. Cham: Springer, pp. 269–304.

Gómez-Jáuregui, V. (2009) Controversial origins of tensegrity. In: Proceedings of the International Association for Shell and Spatial Structures (IASS) Symposium, Valencia Evolution and Trends in Design, Analysis and Construction of Shell and Spatial Structures, 28 September–2 October, pp. 1642–1652. Valencia: Universidad Politecnica de Valencia.

Gómez-Jáuregui, V. (2010) Tensegrity structures and their application to architecture. Santander: PubliCan—Editorial Universidad de Cantabria.

Gómez-Jáuregui, V., Otero, C., Arias, R., Manchado, C. (2012) Novel technique for obtaining double-layer tensegrity grids. International Journal of Space Structures, 27(2–3), 155–166.

Gómez-Jáuregui, V., Otero, C., Arias, R., Manchado, C. (2013) Innovative families of double-layer tensegrity grids: Quastruts and sixstruts. Journal of Structural Engineering, 139(9), 1618–1636.

Gómez-Jáuregui, V., Quilligan, M., Manchado, C., Otero, C. (2018) Design, fabrication and construction of a deployable double-layer tensegrity grid. Structural Engineering International, 28(1), 13–20.

Gough, M.E. (1998) In the laboratory of constructivism: Karl Ioganson's cold structures. October, (84), 90–117 [new version as Gough, M.E. (2005) In the laboratory of constructivism. In: The artist as producer: Russian constructivism in revolution. Berkeley, CA: University of California Press, pp. 60–99].

Gough, M.E. (2005) The artist as producer: Russian constructivism in revolution. Modernism/Modernit. Berkeley, CA: University of California Press.

Grip, R. (1992) The correspondence between convex polydedra and tensegrity systems: A classification system. International Journal of Space Structures, 7(2), 115–125.

Hall, L. (2015) Super ball bot. *National Aeronautics and Space Administration (NASA)*. Available at: www.nasa.gov/content/super-ball-bot

Hanaor, A. (1987) Preliminary investigation of double-layer tensegrities. In: H.V. Topping, ed., Proceedings of international conference on the design and construction of non-conventional structures (Vol. 2). Edinburgh: Civil-Comp Press.

Hanaor, A. (1988) Prestressed pin-jointed structures—flexibility analysis and pre-stress design. Computers and Structures, 28(6), 757–769.

Hanaor, A. (1990) Double-layer tensegrity grids: Geometric configuration and behaviour. In: H. Nooshin, ed., Space structures: Theory and design. London: Multi-Science.

Hanaor, A. (1991a) Double-layer tensegrity grids: Static load response. I—analytical study. Journal of Structural Engineering, 117(6), 1660–1674.

Hanaor, A. (1991b) Double-layer tensegrity grids: Static load response. II—experimental study. Journal of Structural Engineering, 117(6), 1675–1684.

Hanaor, A. (1992) Aspects of design of double-layer tensegrity domes. International Journal of Space Structures, 7(2), 101–114.

Hanaor, A. (1993) Double-layer tensegrity grids as deployable structures. International Journal of Space Structures, 8(1–2), 135–145.

Hanaor, A. (1994) Geometrically rigid double-layer tensegrity grids. International Journal of Space Structures, 9(4), 227–238.

Heartney, E. (2013 [2009]) Kenneth Snelson: Art and ideas. New York: Marlborough Gallery. Available at: http://kennethsnelson.net/KennethSnelson_Art_And_Ideas.pdf

Hrazmi, I., Averseng, J., Quirant, J., Jamin, F. (2021) Deployable double layer tensegrity grid platforms for sea accessibility. Engineering Structures, 231, 111706.

Ingber, D.E. (1993) Cellular tensegrity: Defining new rules of biological design that govern the cytoskeleton. Journal of Cell Science, 107(3), 613–627.

Ingber, D.E. (1998) The architecture of life. *Scientific American Magazine*, January. Available at: http://vv.arts.ucla.edu/projects/ingber/ingber.html

Ingber, D.E. (2003) Tensegrity I. Cell structure and hierarchical systems biology. Journal of Cell Science, (116), 1157–1173. Available at: http://intl-jcs.biologists.org/cgi/content/full/116/7/1157

Intrigila, C., Micheletti, A., Nodargi, N., Artioli, E., Bisegna, P. (2022) Fabrication and experimental characterization of a bistable tensegrity-like unit for lattice metamaterials. Additive Manufacturing, 57, 102946.

Juan, S.H., Tur, J.M.M. (2008) Tensegrity frameworks: Static analysis review. Mechanism and Machine Theory, 43(7), 859–881.

Kanchanasaratool, N., Williamson, D. (2002) Modelling and control of class NSP tensegrity structures. International Journal of Control, 75(2), 123–139.

Kenner, H. (1976) Geodesic math and how to use it (1st ed.). Berkeley, CA: University of California Press.

Kim, K., Moon, D., Bin, J.Y., Agogino, A.M. (2017) Design of a spherical tensegrity robot for dynamic locomotion. In: 2017 IEEE/RSJ International Conference on Intelligent Robots and Systems (IROS), pp. 450–455. Vancouver: IEEE. Available at: https://doi.org/10.1109/IROS.2017.8202192

Kono, Y., Choong, K.K., Shimada, T., Kunieda, H. (2000) An experimental investigation of a type of double layer tensegrity grids. Journal of the International Association for Shell and Spatial Structures (IASS), 41(131).

Korkmaz, S., Bel Hadj Ali, N., Smith, I.F.C. (2011) Determining control strategies for damage tolerance of an active tensegrity structure. Engineering Structures, 33(6), 1930–1939.

Korkmaz, S., Bel Hadj Ali, N., Smith, I.F.C. (2012) Configuration of control system for damage tolerance of a tensegrity bridge. Advanced Engineering Informatics, 26(1), 145–155.

Kurtz, S.A. (1968) Kenneth Snelson: The elegant solution. Art News, October. Available at: www.kennethsnelson.net/icons/art.htm

Lazopoulos, K.A. (2004) Stability of an elastic cytoskeletal tensegrity model. International Journal of Solids and Structures, 42, 3459–3469.

Lazopoulos, K.A., Lazapolou, N.K. (2005) On the elastica solution of a tensegrity structure: Application to cell mechanics. Acta Mechanica, 182, 253–263.

Lee, H., Jang, Y., Choe, J.K., Lee, S., Song, H., Lee, J.P., Lone, N., Kim, J. (2020) 3D-printed programmable tensegrity for soft robotics. Science Robotics, 5(45).

Lee, S., Lee, J. (2014) Form-finding of tensegrity structures with arbitrary strut and cable members. International Journal of Mechanical Sciences, 85, 55–62.

Lessard, S., Castro, D., Asper, W., Chopra, S.D., Baltaxe-Admony, L.B., Teodorescu, M., SunSpiral, V., Agogino, A. (2016) A bio-inspired tensegrity manipulator with multi-DOF, structurally compliant joints. In: 2016 IEEE/RSJ International Conference on Intelligent Robots and Systems (IROS), pp. 5515–5520. New York: IEEE.

Levin, S.M. (2002) The tensegrity-truss as a model for spine mechanics: Biotensegrity. Journal of Mechanics in Medicine and Biology, 2(3–4), 375–388.

Li, Y., Feng, X.Q., Cao, Y.P., Gao, H. (2010) Constructing tensegrity structures from one-bar elementary cells. Proceedings of the Royal Society A: Mathematical, Physical and Engineering Sciences, 466(2010), 45–61. doi:10.1098/rspa.2009.0260

Liu, K., Wu, J., Paulino, G.H., Qi, H.J. (2017) Programmable deployment of tensegrity structures by stimulus-responsive polymers. Scientific Reports, 7, 3511.

Liu, K., Zegard, T., Pratapa, P.P., Paulino, G.H. (2019) Unraveling tensegrity tessellations for metamaterials with tunable stiffness and bandgaps. Journal of the Mechanics and Physics of Solids, 131, 147–166.

Liu, Y., Bi, Q., Yue, X., Wu, J., Yang, B., Li, Y. (2022a) A review on tensegrity structures-based robots. Mechanism and Machine Theory, 168, 104571.

Liu, Y., Dai, X., Wang, Z., Bi, Q., Song, R., Zhao, J., Li, Y. (2022b) A tensegrity-based inchworm-like robot for crawling in pipes with varying diameters. IEEE Robotics and Automation Letters, 7(4), 11553–11560.

Lodder, C. (1992) The transition to constructivism. The great Utopia. In: The Russian and Soviet avant-garde, 1915–1932. New York: Guggenheim Museum.

Ma, S., Chen, M., Peng, Z., Yuan, X., Skelton, R.E. (2022) The equilibrium and form-finding of general tensegrity systems with rigid bodies. Engineering Structures, 266, 114618.

Masic, M., Skelton, R.E., Gill, P.E. (2005) Algebraic tensegrity form-finding. International Journal of Solids and Structures, 42, 4833–4858.

Maxwell, C. (1864) On the calculation of equilibrium and stiffness of frames. Philosophical Magazine, 27, 294–299.

Micheletti, A., dos Santos, F.A., Sittner, P. (2018) Superelastic tensegrities: Matrix formulation and antagonistic actuation. Smart Materials and Structures, 27, 105028.

Micheletti, A., Podio-Guidugli, P. (2022) Seventy years of tensegrities (and counting). Archive of Applied Mechanics, 92, 2525–2548.

Micheletti, A., Williams, W.O. (2007) A marching procedure for form-finding for tensegrity structures. Journal of Mechanics of Materials and Structures, 2, 101–126.

Moholy-Nagy, L. (1929) Von Material zu Architektur. Langen: Bauhausbücher, München.

Motro, R. (1984) Forms and forces in tensegrity systems. In: Proceedings of the Third International Conference on Space Structures, Amsterdam, Holland.

Motro, R. (1987) Tensegrity systems for double layer space structures. In: B.H.V. Topping, ed., Proceedings of the International Conference on the Design and Construction of Non Conventional Structures, Londres, pp. 43–51. London: Multi-Science Publishing Co. Ltd. doi:10.1177/026635119200700201

Motro, R. (1990) Tensegrity systems and geodesic domes. Special Issue of the International Journal of Space Structures, "Geodesic Space Structures", 5(3), 343–354.

Motro, R. (1992) Tensegrity systems: The state of the art. Journal of Space Structures, 7(2), 75–83.

Motro, R. (2003) Tensegrity: Structural systems for the future. London: Kogan Page Science.

Motro, R., Najari, S., Jouanna, P. (1986a) Static and dynamic analysis of tensegrity systems. In: Proceedings of ASCE International Symposium on Shells and Spatial Structures: Computational Aspects. New York: Springer-Verlag.

Motro, R., Najari, S., Jouanna, P. (1986b) Tensegrity systems from design to realisation. In: V. Sedlak, ed., Proceedings of the First International Conference on Lightweight Structures in Architecture, Sydney, pp. 690–697. Sydney: Unisearch Limited.

Nishimura, Y., Murakami, H. (2001) Initial shape finding and modal analysis of cyclic frustum tensegrity modules. Computer Methods in Applied Mechanics and Engineering, 190(43–44), 5795–5818.

Obara, P. (2019) Application of linear six-parameter shell theory to the analysis of orthotropic tensegrity plate-like structures. Journal of Theoretical and Applied Mechanics, 57(1), 167–178.

Obara, P., Tomasik, J. (2020) Parametric analysis of tensegrity plate-like structures: Part 1—qualitative analysis. Applied Sciences, MDPI, 10(20), 7042.

Obara, P., Tomasik, J. (2021) Parametric analysis of tensegrity plate-like structures: Part 2—quantitative analysis. Applied Sciences, MDPI, 11(2), 602.

Olejnikova, T. (2012) Double layer tensegrity grids. Acta Polytechnica Hungarica, 9(5), 95–106.

Pagitz, M., Tur, J.M.M. (2009) Finite element based form-finding algorithm for tensegrity structures. International Journal of Solids and Structures, 46(17), 3235–3240.

Paul, C., Lipson, H., Cuevas, F.J.V. (2005) Design and control of tensegrity robots for locomotion. IEEE Transactions on Robotics, 22(5), 944–957.

Pellegrino, S. (1989) Analysis of prestressed mechanisms. International Journal of Solids and Structures, 26(12), 1329–1350.

Pellegrino, S., Calladine, C.R. (1986) Matrix analysis of statically and kinematically indetermined frameworks. International Journal of Solids and Structures, 22(4), 409–428.

Pugh, A. (1976) An introduction to tensegrity. Berkeley, CA: University of California Press.

Quilligan, M., Gómez-Jáuregui, V., Manchado, C., Otero, C. (2020) Development and testing of a deployable double layer tensegrity grid. In: Proceedings of Civil Engineering Research in Ireland, Cork Institute of Technology, August.

Rieffel, J., Valero-Cuevas, F., Lipson, H. (2009) Automated discovery and optimization of large irregular tensegrity structures. Computers & Structures, 87(5–6), 368–379.

Riether, G., Wit, A.J. (2016) Underwood pavilion: A parametric tensegrity structure. In: Advances in architectural geometry, vdf Hochschulverlag AG an der ETH Zürich, Zürich, Switzerland, pp. 188–203. doi: 10.3218/3778-4_14

Rimoli, J.J. (2018) A reduced-order model for the dynamic and post-buckling behavior of tensegrity structures. Mechanics of Materials, 116, 146–157.

Rimoli, J.J., Pal, R.K. (2017) Mechanical response of 3-dimensional tensegrity lattices. Composites Part B: Engineering, 115, 30–42.

Roth, B., Whiteley, W. (1981) Tensegrity frameworks. Transactions of the American Mathematical Society, 265(2), 419–446.

Rovira, A.G., Mirats Tur, J.M. (2009) Control and simulation of a tensegrity-based mobile robot. Robotics and Autonomous Systems, 57(5), 526–535.

Sabelhaus, A.P., Bruce, J., Caluwaerts, K., Manovi, P., Firoozi, R.F., Dobi, S., Agogino, A.M., SunSpiral, V. (2015) System design and locomotion of SUPERball, an untethered tensegrity robot. In: 2015 IEEE International Conference on Robotics and Automation (ICRA), pp. 2867–2873. Seattle, WA: IEEE.

Sadao, S. (1996) "Fuller on Tensegrity", in origins of tensegrity: Views of Emmerich, Fuller and Snelson. International Journal of Space Structures, 1(1–2), 37–42.

Salahshoor, H., Pal, R.K., Rimoli, J.J. (2018) Material symmetry phase transitions in three-dimensional tensegrity metamaterials. Journal of the Mechanics and Physics of Solids, 119, 382–399.

Salerno, G. (1992) How to recognize the order of infinitesimal mechanisms: A numerical approach. International Journal for Numerical Methods in Engineering, 35(7), 1351–1395. doi:10.1002/nme.1620350702

Schek, H.J. (1974) The force density method for form finding and computation of general networks. Computer Methods in Applied Mechanics and Engineering, 3(1), 115–134.

Schlaich, M. (2004) The Messeturm in Rostock: A tensegrity tower. Journal of the International Association for Shell and Spatial Structures (IASS), 45(145), 93–98.

Schneider, A. (1977) Interview with Kenneth Snelson. Nationalgalerie Berlin Exhibition Catalog, March–May. Available at: www.kennethsnelson.net/icons/art.htm

Schorr, P., Li, E.R.C., Kaufhold, T., Hernández, J.A.R., Zentner, L., Zimmermann, K., Böhm, V. (2021) Kinematic analysis of a rolling tensegrity structure with spatially curved members. Meccanica, 56, 953–961.

Scolamiero, L., Zolesi, V., Ganga, P.L., Podio-Guidugli, P., Micheletti, A., Tibert, G. (2015) A deployable tensegrity structure, especially for space applications. European Patent No. 2828928.

Scolamiero, L., Zolesi, V., Ganga, P.L., Podio-Guidugli, P., Micheletti, A., Tibert, G. (2017) A deployable tensegrity structure, especially for space applications. U.S. Patent No. 9815574.

Scruggs, J.T., Skelton, R.E. (2006) Regenerative tensegrity structures for energy harvesting applications. In: Proceedings of the Conference on Decision and Control, San Diego, CA.

Shibata, M., Hirai, S. (2009) Rolling locomotion of deformable tensegrity structure. Mobile Robotics, 479–486.

Skelton, R.E., de Oliveira, M.C. (2009) Tensegrity systems. New York: Springer.

Smaili, A.E., Motro, R., Raducanu, V. (2004) New concept for deployable tensegrity systems, structural mechanics activated by shear force. In: Proceedings of IASS2004, Montpellier.

Snelson, K.D. (1965) Continuous tension, discontinuous compression structures. U.S. Patent No. 3,169,611, 16 February.

Snelson, K.D. (1990) "Letter from Kenneth Snelson" to R. Motro, International Journal of Space Structures, Space Structures Research Centre, Department of Civil Engineering, University of Surrey, Guildford, Surrey, November 1990. Available at: https://www.grunch.net/snelson/rmoto.html

Snelson, K.D. (1996) Snelson on the tensegrity invention. International Journal of Space Structures, 11(1–2), 43–48.

Snelson, K.D. (2003) Letter from Kenneth Snelson to Maria Gough on Karl Ioganson, 17 June. Available at: https://bobwb.tripod.com/synergetics/photos/snelson_gough.html Accessed June 2022.

Snelson, K.D. (2012) The art of tensegrity. International Journal of Space Structures, 27(2–3), 71–80.

Stadium Guide (2022) Estadio Único Ciudad de la Plata. Available at: www.stadiumguide.com/ciudaddelaplata/

Stern, I. (2003) Deployable reflector antenna with tensegrity support architecture and associated methods. U.S. Patent No. 6542132.

Sultan, C. (2009) Tensegrity: 60 years of art, science, and engineering. In: Advances in applied mechanics (Vol. 43). Burlington: Academic Press, pp. 69–145.

Sultan, C., Corless, M., Skelton, R.E. (2001) The prestressability problem of tensegrity structures. Some analytical solutions. International Journal of Solids and Structures, 38–39, 5223–5252.

Sultan, C., Corless, M., Skelton, R.E. (2002a) Symmetrical reconfiguration of tensegrity structures. International Journal of Solids and Structures, 39(8), 2215–2234.

Sultan, C., Corless, M., Skelton, R.E. (2002b) Linear dynamics of tensegrity structures. Engineering Structures, 26(6), 671–685.

Sultan, C., Skelton, R.E. (1998a) Force and torque smart tensegrity sensor. In: V.V. Varadan, ed., Proceedings of SPIE symposium on smart structures and materials (Vol. 3323). San Diego, CA: V.V. Varadan. Available at: https://doi.org/10.1117/12.316316

Sultan, C., Skelton, R.E. (1998b) Tendon control deployment of tensegrity structures. In: V.V. Varadan, ed., Proceedings of SPIE, smart structures and materials: Mathematics and control in smart structures (Vol. 3323), pp. 455–466. San Diego, CA: V.V. Varadan.

Sultan, C., Skelton, R.E. (2003a) Deployment of tensegrity structures. International Journal of Solids and Structures, 40(18), 4637–4657.

Sultan, C., Skelton, R.E. (2003b) Tensegrity structures prestressability investigation. International Journal of Space Structures, 18(1), 15–30.

Sultan, C., Skelton, R.E. (2004) A force and torque tensegrity sensor. Sensors and Actuators Journal A: Physical, 112(2–3), 220–231.

Sultan, C., Stamenovic, D., Ingber, D. E. (2004) A computational tensegrity model explains dynamic rheological behaviors of living cells. Annals of Biomedical Engineering, 32(4), 520–530.

Tarnai, T. (1980) Simultaneous static and kinematic indeterminacy of space trusses with cyclic symmetry. International Journal of Solids and Structures, 16(4), 347–359.

TensiNet (2022) Seoul Olympic gymnastics hall and fencing hall. Available at: www.tensinet.com/index.php/component/tensinet/?view=project&id=4013

Tibert, A.G., Pellegrino, S. (2002) Deployable tensegrity reflector for small satellites. Journal of Spacecraft and Rockets, 39, 701–709.

Tibert, A.G., Pellegrino, S. (2003) Review of form-finding methods for tensegrity structures. International Journal of Space Structures, 18(4), 209–223.

Tibert, G. (2002) Deployable tensegrity structures for space applications. Doctoral Thesis, Royal Institute of Technology, Department of Mechanics Stockholm. Available at: http://www-civ.eng.cam.ac.uk/dsl/publications/TibertDocThesis.pdf

Tran, H.C., Lee, J. (2010) Advanced form-finding of tensegrity structures. Computers & Structures, 88(3–4), 237–246.

Tran, H.C., Lee, J. (2013) Form-finding of tensegrity structures using double singular value decomposition. Engineering with Computers, 29, 71–86.

Uitz, B. (1922) Egyseg, No. 2, Vienna, June 1992; reproduced in "The first Russian show: A commemoration of the Van Diemen exhibition. Berlin". London: Annely Juda Fine Art, 1983, p. 41.

Valero-Cuevas, F.J., Yi, J.W., Brown, D., Namara, R.V.M., Paul, C., Lipson, H. (2007) The tendon network of the fingers performs anatomical computation at a macroscopic scale. IEEE Transactions on Biomedical Engineering, 54(6), 1161–1166.

Vangelatos, Z., Micheletti, A., Grigoropoulos, C.P., Fraternali, F. (2020) Design and testing of bistable lattices with tensegrity architecture and nanoscale features fabricated by multiphoton lithography. Nanomaterials, 10(4), 652.

Vassart, N., Motro, R. (1999) Multiparametered form-finding method: Application to tensegrity systems. International Journal of Space Structures, 14(2), 147–154.

Vilnay, O. (1990) Cable nets and tensegric shells: Analysis and design applications. New York: Ellis Horwood.

Volokh, K., Vilnay, O., Belsky, M. (2000) Tensegrity architecture explains linear stiffening and predicts softening of living cells. Journal of Biomechanics, 33(12), 1543–1549.

Wang, B.B. (1998) Cable-strut systems: Part I—tensegrity. Journal of Constructional Steel Research, 45(3), 281–289.

Wang, B.B. (2004) Free-standing tension structures: From tensegrity systems to cable-strut systems. London; New York: Spon Press.

Wang, B.-B., Li, Y.Y. (1998) Definition of tensegrity systems. Can dispute be settled? In: Richard Hough and Robert Melchers (Eds.), Proceedings of LSA98 "Lightweight Structures in Architectural Engineering and Construction" (Vol. 2), pp. 713–719. Sydney: The Lightweight Structures Association of Australasia (LSAA) Inc.

Wang, B.-B., Li, Y.Y. (2003) Novel cable-strut grids made of prisms: Part I. Basic theory and design. International Journal of Space Structures, 44(142), 93–125.

Wang, N., Naruse, K., Stamenovic, D., Fredberg, J.J., Mijailovich, S.M., Tolic-Nørrelykke, I.M., Polte, T., Mannix, R., Ingber, D.E. (2001) Mechanical behavior in living cells consistent with the tensegrity model. Biophysics and Computational Biology, 98(14), 7765–7770.

Wendling, S., Canadas, P.V., Chabrand, P. (2003) Towards a generalized tensegrity model describing the mechanical behaviour of the cytoskeleton structure. Computer Methods in Biomechanics and Biomedical Engineering, 6(11), 45–52.

Xu, X., Luo, Y. (2010) Form-finding of nonregular tensegrities using a genetic algorithm. Mechanics Research Communications, 37(1), 85–91.

Xu, X., Wang, Y., Luo, Y. (2016) General approach for topology-finding of tensegrity structures. Journal of Structural Engineering, ASCE (American Society of Civil Engineers), 142(10), 04016061.

Zappetti, D., Arandes, R., Ajanic, E., Floreano, D. (2020) Variable-stiffness tensegrity spine. Smart Materials and Structures, 29(7), 075013.

Zappetti, D., Mintchev, S., Shintake, J., Floreano, D. (2017) Bio-inspired tensegrity soft modular robots. In: Biomimetic and Biohybrid Systems. Cham: Springer International Publishing, pp. 497–508.

Zawadzki, A., Al Sabouni-Zawadzka, A. (2020) In search of lightweight deployable tensegrity columns. Applied Sciences, MDPI, 10(23), 8676.

Zhang, J.Y., Ohsaki, M. (2006) Adaptive force density method for form-finding problem of tensegrity structures. International Journal of Solids and Structures, 43, 5658–5673.

Zhang, L., Maurin, B., Motro, R. (2006) Form-finding of nonregular tensegrity systems. Journal of Structural Engineering, 132(9), 1435–1440.

Zolesi, V.S., Ganga, P.L., Scolamiero, L., Micheletti, A., Podio-Guidugli, P., Tibert, G., Donati, A., Ghiozzi, M. (2012) On an innovative deployment concept for large space structures. In: 42nd International Conference on Environmental Systems, San Diego, CA, 15–19 July. San Diego, CA: ICES Organization.

Index

Note: Page numbers in *italics* indicate a figure on the corresponding page.

For Product Safety Concerns and Information please contact our EU
representative GPSR@taylorandfrancis.com
Taylor & Francis Verlag GmbH, Kaufingerstraße 24, 80331 München, Germany

www.ingramcontent.com/pod-product-compliance
Lightning Source LLC
Chambersburg PA
CBHW060406220326
41598CB00023B/3037

9 781032 440361